THE DIALECTICAL BEHAVIOR THERAPY SKILLS

The Most Powerful Strategies to Overcome Anxiety by Learning How to Manage Your Emotions, Reduce Symptoms, and Get Back to Your Life

Behavioral EQ Institute

Table of Contents

Introduction

The challenge for someone who desires to recover his life but hasn't established a constructive attitude is that there will be a series of usual negative thoughts and reactions which arise on a justly consistent basis. These psychological propensities should be tended to so as to improve things - negative reasoning must be transmuted into something which can profit you. Typically, our pessimism originates from off base reasoning examples. For instance, we may make clearing speculations and arrive at a general determination dependent on a solitary occasion or one bit of proof. Whenever something terrible happens once, we anticipate that it should happen over and over. Such contemplations frequently incorporate the words "consistently" and "never" such "I neglected to complete that task on schedule. I never accomplish things right".

Seeing the exercise scholarly and self-recuperating inside and remain on the way of inspirational viewpoint. Downplaying certain circumstances will enable you to feel and think increasingly more grounded and emphatically about the hindrances in life you face. To conquer issues and obstructions in life, you initially should not be so difficult on yourself and take a gander at your core belief with persistence. Each time you endeavor to beat issues, you find new things about yourself and you venture up to a more significant level of recognition and experience expanded inward quality.

This must reveal to us that since the affection we have inside us is genuine, and along these lines must be of

God, at that point this adoration must not have a solitary exemption. What I mean here is that we make up our own special cases, similarly as we make up our very own convictions. A sentiment of needing to surrender or an inclination that a circumstance is difficult to get out is one of the most noticeably awful sorts of emotions you can ever have. In any case, it's a piece of the range of human feelings, and it's something that everybody has encountered eventually, and will likely involvement later on.

The fact of the matter is there are good and bad times in each way. There are high points and low points to your life at the present time, and there will be good and bad times on your street to progress. You can't hope to never again feel along these lines... Be that as it may, you can manage it much superior to anything you may have before.

Acknowledge the sentiments you are having, yet perceive and discharge the negative contemplations. Musings like "This is inconceivable" and "I should surrender" and "there is no expectation." Just let go of these considerations, over and over. In the event that they come up, enable them to emerge, and afterwards, let go of them and refocus on the current circumstance. A decent mantra to help you to remember this - "Think about the arrangement, not the issue."

Your state of mind won't change in a split second, however, realize that by playing out this procedure over and over, you will routinely authorize yourself to act along these lines. Do you know somebody who regardless of what circumstance they are in, regardless of how terrible things get, they never whine and scarcely even show they are experiencing difficulty? That is not on the grounds that they don't feel any thrashing whatsoever - it's since they routinely relinquished the negative contemplations which emerge. They are pondering the arrangement, as opposed to the destruction.

A typical propensity for adverse individuals, and by pessimistic individuals, I mean a tremendous extent of the populace, is to gripe about things when they are not how they need them to be. What number of individuals do you know who routinely cry about their position at home and at work? What number of individuals grumble about their accomplice, their vehicle, their home, their life, etc.?

Whining accomplishes next to no for you. A certified grumbling about low-quality assistance so as to expect a discount is a certain something. However, it is totally silly to gripe and cry about things like your activity, your circumstance, your life. There is just a single thing this activity unquestionably accomplishes; it makes individuals consider you a complainer. Regardless of whether they don't state it, or deliberately think it, constructive individuals will spot it a mile off and nobody likes to associate with it.

Whenever you want to gripe, don't utter a word. Acknowledge the idea, let it go, and afterwards think "what would I be able to do to improve this?" The appropriate responses may astound you. Maybe it's something as basic as changing an easily overlooked detail anywhere, or possibly you have to change something major... perhaps you'll find that you have to get another line of work, possibly you'll find that you have to have a significant chat with your accomplice about something which has been annoying you. Whatever the circumstance ends up being - think about the arrangement and make a positive move to improve it.

What is DBT?

Dialectical Behavior Therapy is a proof-based cognitive-behavioral treatment created by Dr. Marsha Linehan. DBT has been proved to be robust for people battling with directing feelings and tolerating distress. The objective of DBT is to enable patients to figure out how to deal with their painful emotions by allowing them to experience, perceive, and accept them. As you learn to acknowledge and manage your feelings, you can supplant maladaptive methods for dealing with stress with healthy/versatile adapting skill. To enable you to accomplish this, DBT advisors use an equalization of acceptance and change strategies.

Dialectical Behavior Therapy or DBT was initially created by Marsha Linehan in the late 1980s as an approach to treat and help deal with the symptoms of Borderline Personality Disorder.

The Dialectical part, or the "D" in DBT, is a Buddhist idea of contradicting powers. For instance, somebody who was sexually abused by a parent may feel both loves and loathe for that parent simultaneously. That is a hard division to experience, and the moving part instructs us to sit tranquility with those warring powers within our brains.

The Behavior Therapy ("BT") alludes to the use of standards of Cognitive Behavioral Therapy, or CBT, for example, the possibility that we as a whole have thoughts that cause us to have feelings, which at that point cause us to follow up on those emotions. Take, for example, when somebody cuts you off in traffic, and you think, "That individual place me in risk (and different interjections)!" That idea causes you to be furious and frightened, which at that point causes you to blare your horn or demonstrate that driver one of your fingers.

Components Of DBT

There are four components of full DBT: skill training group, singular treatment, DBT telephone instructing, and consultation group.

DBT abilities preparing group is focused on improving clients' capacities by showing them behavioral skills. The group is run like a class where the group head teaches the skills and assigns schoolwork for clients to work on utilizing the skills in their regular day to day existences. Groups meet on week by week basis for around 2.5 hours, and it takes 24 weeks to get past the full skill educational program, which is regularly rehashed to make a 1-year program. Briefer timetables that show just a subset of the skills have also been produced for specific populaces and settings.

DBT singular therapy is focused on improving clients inspiration and helping clients to apply the skills to specific challenges and occasions in their lives. In the standard DBT example, singular therapy happens once per week for around an hour and runs simultaneously with skill groups.

DBT telephone instructing is focused on giving clients in-the-minute training on how to use skill to successfully adapt to stressful situations that emerge in their regular day to day existences. Clients can call their specialist between sessions to get training at the occasions when they need assistance the most.

DBT therapist meeting group is expected to be therapy for the therapists and to help DBT suppliers in their work with people who frequently have a serious, intricate, difficult-to-treat issue. The counsel group is intended to enable specialists to remain persuaded and skilled so they can give the ideal treatment. Groups typically meet weekly and are made out of individual therapists and group pioneers who offer obligation regarding every customer's consideration.

Dialectical Behavior Therapy Self Help

There are many things you can do on your own when it comes to practicing dialectical behavior therapy strategies. Many people who are not struggling with any mental health issue disorder decide to embrace those DBT strategies and techniques to boost their mental health, to bring more joy and happiness into their lives and to simply feel better.

This is where dialectical behavior therapy steps in with its highly valuable skills and strategies that everyone can benefit from. In fact, those dialectical behavior therapy strategies which patients learn during their treatments are very valuable to everyone.

Skills to Embrace

There are three major skills which patients learn during their dialectical behavior therapy sessions, you can embrace as well.

Today, DBT is used for treating patients struggling with many different mental health disorders while its valuable strategies can be of great importance to all people. DBT teaches valuable behavioral skills including interpersonal effectiveness, emotion regulation, distress tolerance and mindfulness which can be beneficial to everyone.

No matter if you have a mental health disorder or not, you can take these valuable lessons, incorporate them into your life and enjoy benefits which they bring.

As you embrace these valuable lessons, you get to enjoy a much healthier and happier lifestyle with almost no negative emotions to distract you.

As you learn how to properly manage your emotions, you get to take control over your life fully, you get to build more meaningful relationships, be more present in all social environments and simply enjoy the life as you should.

As patients learn at their DBT sessions, the first thing you need to embrace is mindfulness. This means you must live your life more in the current moment instead of thinking about those past things and instead of anticipating what may happen in the future.

You should not let yourself to be hijacked by the future or by the past. As you practice mindfulness, you will be more aware of your surroundings, of people around you. You will also be more aware of your reactions, actions, feelings, and thought.

Practicing mindfulness also means that you gain control over your emotions, so you can check in, pause as well as analyze and identify all your emotions, so you can easily make those healthy decisions.

To practice this valuable dialectical behavior therapy skills, you should start living mindfully. This means, when walking you should try to feel your body as it walks slowly.

Make sure you pay attention to getting to know your body, its needs, pay attention to all those muscles used for walking, pay attention how they move just to achieve that goal of walking.

While walking, make sure you pay attention to all those things and all people surrounding you, look at the sky, notice all different colors, notice the trees around you, pay attention to your neighborhood and everything else. Make sure, you notice everything, everything beautiful happening around you.

In those cases when your mind starts wandering, make sure you redirect it to the current moment. Make sure you focus on those external experiences again and start again noticing what is happening around you. You can also focus on your internal experiences, your feelings, thoughts, various physical sensations.

In those cases when you feel as you are entangled in your feelings and emotions, make sure you focus more on those positive emotions which bring joy and happiness rather than thinking and observing those negativities.

Another valuable lesson patients learn at their dialectical behavior therapy sessions is reality acceptance. This is one of the main skills you should embrace when it comes to your daily experiences, emotions, reactions or feelings.

This skill focuses on accepting everything that is happening around you, no matter its intensity, no matter being positive or negative. If you find accepting reality as something harsh, you probably want to avoid thinking about it, however, this approach only makes it worse.

Instead, you should work on accepting those negative emotions, those negative, often painful events rather than trying to forget about them or to avoid them. In fact, if you fight reality, it just makes your suffering worse.

For instance, if you are somewhere alone, sitting bored, you probably start thinking what happened recently which brought some negativity in your life. You probably start thinking about other more positive things which could have happened, but they did not.

Thinking about all the things which could have been different is not going to help you, just worsen your situation, so the best approach and the best thing you can do for yourself is to accept those painful emotions or events, rather than avoiding them or trying to forget that they ever happened.

Another valuable skill which patients learn at DBT sessions is nonjudgmental stance which focuses on teaching people to be less judgmental when it comes to everything, people, situations, etc.

When you start judging people or things as good or bad, you should stop immediately. Those negative judgments you have tend to boost your own negativity bringing more emotional pain even if you do not notice it at the very beginning.

Therefore, when you feel irritated, angry or frustrated, make sure you pay attention what made you feel that way or which judgment made you feel that way. As you analyze your emotions in this manner, you should focus your efforts on replacing those negative judgments with more positive, with some fact and with some judgment which will bring you more positive emotions.

Instead of thinking that the weather is awful these days, you should say that you are irritated as you have to take a walk while it is raining.

Instead of thinking that you are not a good friend, you should say to yourself that you will improve, and try not to hurt anyone. Instead of saying to your friends that he or she is not a good friend, say that he or she has been there for you many times in the past.

By being less judgmental of a situation, people or everything else around you will not eliminate your emotional pain, but it will help you reduce those negative emotions such as frustration, anger or irritation.

As you embrace this approach, you will be able to think more wisely, clearly as you open multiple choices for yourself. This approach will also empower you to focus on healthy problem-solving approach making a healthy decision which will support and serve you good.

In fact, all us can benefit from working on managing our emotions, from becoming more aware of our feelings, thoughts while we accept what is happening in the present moment as we become less judgmental of both other people and ourselves. These skills that dialectical behavior therapy teaches us, in fact, lead to significantly healthier, happier and more joyful life.

Benefit From DBT

DBT treatments in general with many different parts and with different in-depth skills can be extremely intense.

However, as you practice DBT skills more, as you apply them more daily, this way of living becomes something you can benefit from seriously leading to a healthier as well as happier lifestyle.

DBT was developed to help people struggling with different mental health disorders. However, it can provide valuable lessons to everyone who decides to embrace DBT strategies and practices. Whether you are feeling unhappy at the moment or you feel as you are going through some very intense crisis, DBT can help you.

There are several easy ways you can start using DBT practices on your own. You can start with turning off the music during your daily commutes. You listening to the music as you take your regular commutes is opposite to those DBT strategies and practices.

This kind of distraction can be a lot of fun, but as you enjoy it, you miss all those things happening around you.

Moreover, you listening to the music during your daily commutes may lead to missing something important such as a public service announcement telling something important that you are not able to hear due to music. The most important thing here is that listening to music in these situations keeps you away from being mindful of everything which is happening around you.

It also cuts you from those things going on around you. This is something opposite to the DBT strategies which teach you to be more mindful, be present in the current moments.

As you start working on being mindful in those small areas, in some time, you will be able to be more mindful in some bigger areas which matter the most. To kick off this journey, you can start by turning off music during your commute.

Another valuable lesson we have learned from those DBT strategies and practices is analyzing and identifying both positive and negative emotions. The best approach for this is to practice naming your emotions in those everyday conversations.

Everyone feels sometimes angry, frustrated or upset and sometimes these emotions come from nowhere for no apparent reason. In these cases, you simply do not know why you feel angry.

This is where dialectical behavior therapy steps in. It encourages us to improve our ability to be aware of all those emotions both positive and negative.

It encourages us to be aware of our reactions, feelings, and thoughts in a manner that empowers us to be more capable of managing them. One of the best practices you can embrace is to start naming your emotions in your daily conversations.

As you do so, another valuable lesson is to always listen to the person you are interacting with. As you do so, you will accomplish your goals when it comes to dealing with someone meaning you will improve your interpersonal effectiveness which is another valuable skill taught by dialectical behavior therapy.

As you embrace this lesson, you will also increase the likelihood that other people feel heard, so there is a higher chance that you will get what you want. This approach also significantly increases the overall effectiveness of your possible response, as your choice is based on some more informed assessment of that situation.

And, as you embrace this DBT lessons, make sure you always ask yourself what you really want, what you want to accomplish and which lifestyle you truly want to lead.

- **Wise Mind Accepts**

Wise Mind is one of three mindsets that are instructed in DBT. This is the perfect perspective where you can consolidate your reasonable considerations with your emotions so you can take an increasingly balanced and natural way to deal with perplexing situations.

Below is a practical DBT tool that will enable you to build up the ability of diversion with the goal that you can work in Wise Mind all the more now and again. Indeed, when mastered, you can make this strategy a part of your regular day to day existence. The abbreviation Wise Mind Accepts causes you to recall how to use progressively fun exercises to shift stressful and upsetting thoughts.

Distract with Activities

Distract yourself from upsetting contemplations by taking part in specific exercises that catch your eye. Consider things you would do if you weren't so bothered, and then do them. Have a go at heading off to the exercise center, finishing a crossword puzzle, watching a movie, going through your storeroom to discover things to provide for philanthropy, going for a stroll, or participating in a leisure activity that you cherish. It doesn't make a difference what you do as long as it is a valuable movement, and it gets your mind off of the negative considerations. Also, remember that if you pick a campaign that includes other individuals, shun discussing the situation. The thought is to use disturbing activities to get your mind off of the negative thoughts.

Distract With Donating

Distract yourself by donating to the lives of others. Being in support of other individuals will, in general, be much all the more diverting and accommodating at that point doing exercises for yourself. Contributions can incorporate volunteering for an association, keeping an eye on a companion, cooking a feast for an old individual, helping your sister paint her lounge room, or whatever else that is in support of another person that gets your mind off of your disappointments.

Distract With Comparing

Distract yourself by considering people who are more regrettable off than you. Then again, you can find when things were more terrible in your own life. This can enable you to experience enormous gratitude for your present reality. The fact of the matter isn't to discredit how you feel but to allow you to put your situation into point of view. The way toward looking at advises you that there are situations that are more terrible than yours.

Distract With Inverse Emotions

Participate in exercises that will inspire emotions that are the direct inverse of what you are right now experiencing. For instance, when you're irate, you could take a stab at popping a parody into the DVD player and spend the following couple of hours laughing. Try singing or moving when you're feeling dismal. Or on the other hand, possibly watch somebody sky diving or accomplishing something brave when you're feeling terrified or anxious. The thought is to occupy yourself with exercises that will urge you to experience the opposite emotion.

Distract With Pushing Away

This system allows you to drive negative considerations crazy by mentally leaving the situation. You ought to let yourself to recognize that the issue exists but then intentionally manage it sometime in the future. This isn't an evasion of the significant problems, but it allows you to deliberately separate to experience some transitory help. A portion of the ways that you can adequately execute this method is to envision a fanciful wall among you and the situation. Visualize that the wall isolates you from the cases. You could also take a stab at imagining that you are pushing the volume button on the case with the goal that it is getting lower and lower until you can never again hear it. You could also see yourself putting the issue in a cabinet in another room where it can securely remain until you're in a superior position to manage it. This system allows you fleeting help with the goal that you can pass through any quick emergency or emotional distress.

Distract With Thoughts

Try entertaining your consideration with neutral thoughts that are not linked to your situation. When you fill your head with other evidence, then there's no room left for the negative deliberations. You can give tallying the tile a shot the floor, imagining new furniture for your home, speculating the most loved shade of each kid that passes by in the gym or doing some math estimates in your mind. This will enable you to keep your thoughts off of painful conditions.

Distract With Sensations

You can also occupy yourself using physical and substantial sensations. A few exercises you could try incorporate putting an ice pack on your neck, washing up, drinking a warm refreshment, tuning in to loud music, snapping an elastic band on your wrist, agreeing on a fan to blow in your face, or swimming in freezing water. Exposing yourself to different physical sensations will enable you to activate your body so your mind and emotions will accompany it. Strong impressions allow you to disengage from the agony of your present situation. This is probably the ideal approaches to get "unstuck" when in an emergency.

Wise Minds Accepts a methodology that numerous individuals find very useful during regular day to day existence. It allows you to adopt an increasingly adjusted strategy to upsetting situations. This methodology shows you how to use diversion when confronting situations that are out of your control. Above all, in any case, is that this vital tool can support you to decrease disturbing emotions while at the same time opposing unhelpful and potentially damaging behavior. You can get familiar with Wise Mind ACCEPTS and other DBT methodologies by reaching a Therapist who has some expertise in Dialectical Behavior Therapy. A fast web search will furnish you with a list of incredible clinicians in your general vicinity.

- **DBT Stages and Goals**

When it comes to dialectical behavior therapy stages, there are most commonly four stages included. In stage one, therapists treat those the most self-destructive behaviors in patients such as self-injury or suicide attempts.

In stage two, therapists work with patients encouraging them to address those quality life skills such as distress tolerance, emotional regulation, and interpersonal effectiveness while in stage three, patients need to focus on improving their self-esteem and their relationships. During the stage four, patients work on bringing more relationship connection and more joy into their environments.

Dialectical behavior therapy, in fact, was designed to both comprehensively and systematically treat patients struggling with major mental health issues.

Because dialectical behavior therapy was originally intended for people struggling with suicidal thoughts and extreme emotional distress, DBT treatments come in four different stages for the therapist to address all the concerns.

First Stage

Stage one focuses mainly on the stabilization. During this stage, patients learn how to properly deal with those extreme, sudden emotions as well as with other issues related to their conditions. Patients in this stage most commonly report as they are in a very low point in their lives.

Therefore, the main goal of this stage is to provide patients safety and that first line crisis intervention. The goal is also to help them achieve that greater control over their problematic behaviors.

Second Stage

During the stage two, those extreme patients' behaviors tend to stabilize, but their major symptoms related to their condition may still be present. At this stage, those emotional pains are also most commonly brought to the surface while their traumatic experiences are explored safely.

The main goal of this stage is to encourage patients to fully experience all their emotional pains and traumas, rather than burying or silencing them.

Third Stage

The third stage mainly focuses on enhancing the overall quality of patients' lives through reasonable goal-setting, through life maintenance of progress.

The main challenge here for the patients is to learn to live while building self-respect, building life goals while finding happiness and peace. The goal here is to provide patients with strategies on how to live a life of both ordinary unhappiness and happiness.

Fourth Stage

During the stage four, therapists focus on providing support for patients for them to advance their lives to the following levels. During their time at DBT therapy sessions, patients commonly improve their newly learned skills, learn how to apply them properly to different environments.

During this last stage, therapists also work with their patients on helping them to achieve as well as to maintain that ongoing capacity for success and happiness.

The Efficacy Of Dbt

Finding a wide range of different studies have shown that dialectical behavior therapy is extremely effective especially when used for treating patients struggling with the borderline personality disorder, suicidality, post-traumatic stress disorder, and self-harm.

Dialectical behavior therapy is one of several therapies using evidence-based treatments. Recent studies have shown that DBT is highly effective when it comes to reducing both suicidal and non-suicidal self-injury, suicidal behavior, substance abuse, depression and anxiety symptoms, psychiatric hospitalization dropout. The studies have shown that DBT also improves global and social functioning in patients.

Moreover, multiple independent studies and controlled trials on the effectiveness of DBT have found that patients who underwent DBT sessions for one year showed significantly increased improvement from self-harming behaviors, depression and anxiety symptoms than patients who received some other kind of therapy.

Moreover, multiple studies have shown that dialectical behavior therapy is highly effective when it comes to reducing suicide attempts as patients who received DBT treatments are half likely to have suicidal thoughts as well as to attempt suicide. They also have less psychiatric hospitalization in comparison to patients who receive some other treatments.

Understanding Anxiety

Anxiety in and of itself is not a problematic emotion, as virtually everyone experiences some degree of anxiety in their lives. You may feel anxious about being late to work, or about having a first date with someone that you are really attracted to, and this anxiety could even lead to symptoms like worry, shakiness, or sweating. In situations like these, anxiety is fairly common and is not considered to be problematic or anything to worry about. However, not everyone experiences healthy levels of anxiety when their anxious responses are being triggered by the world around them. Instead, some people experience intense and overwhelming reactions that essentially hijack their minds and prevent them from rationally or reasonably working their way through their anxious experiences.

When you begin experiencing clinical anxiety, the natural symptoms of your anxiety begin to increase significantly. Instead of just experiencing a heightened but still manageable sense of anxiety, your symptoms begin to become unmanageable and overwhelming. People who are experiencing anxiety disorder are believed to have an overactive fight or flight response, which can trigger excessive responses to their original trigger or stressor. For example, the idea of taking a bus by themselves through the city they've lived in all their lives may trigger a full-blown panic attack if they are living with an anxiety disorder. Alternatively, the idea of going to a busy public place might trigger intense feelings of discomfort and stress such as nausea, high blood pressure, and intrusive thoughts. Anxiety disorder is not always characterized by panic attacks, but also by how intense the anxious response to a trigger is, and how easy (or not) it is for the person suffering to regain control over their symptoms.

The understanding as to why anxiety develops this way in certain people is not entirely understood, though there are many reasons as to why certain people will experience anxiety. For example, if a person experiences something that was particularly traumatic or stressful, their anxious response to triggers that stimulate the memory of those challenging experiences might cause anxiety. This type of stimuli can also trigger PTSD, so it is important that if this is what has caused anxiety, you are clear as to whether it is truly anxiety that is being dealt with or if it has progressed to PTSD.

It is also believed that consistent exposure to stress and overwhelm can cause anxiety in people, which can escalate to problematic anxiety over time if they are not able to relax their minds. This likely happens from consistent exposure to cortisol and adrenaline, which are the two hormones responsible for generating feelings of stress and anxiety in people who are experiencing them. Alternatively, being in a volatile relationship with someone can also result in someone experiencing anxiety as they may be trained by their abuser to live "on edge" all the time. This anxiety is used to the advantage of the abuser who relies on their victim to be anxious all the time so that they can easily swing them off balance and abuse them further, without having the anxious victim fight back.

There are truly many reasons as to why a person may develop anxiety, but regardless of how it has developed, problematic anxiety can be troublesome and challenging to ward off. It is believed that more than 40 million people live with anxiety worldwide. Living with anxiety can be life-changing, as it can cause negative consequences in virtually every area of your life if it becomes overwhelming or out of control. For that reason, it is important that anyone who is experiencing anxiety seeks support in dealing with their symptoms so that they can hopefully be treated and resume living a normal life.

How Anxiety Affects People Who Live with It

Anxiety impacts people in different ways, which is why it can be so challenging to deal with as not every case is the same. The symptoms of anxiety can vary depending on each individual case, so identifying anxiety in yourself and knowing when it has become problematic ultimately depends on your personal feeling towards your symptoms. If you feel like you truly cannot get them under control by yourself or like they are negatively impacting your quality of life, chances are, you are dealing with problematic anxiety that needs to be treated properly.

If you are experiencing overwhelming anxiety on an ongoing basis, you might find that you are constantly feeling restless and overwhelmed even when there are no real reasons for you to feel anxious. This feeling of restlessness may come and go randomly for no apparent reason, or it may be constant if you are dealing with a more advanced case of problematic anxiety disorder. When you feel this restlessness, it may feel like every little thing can lead to you having a full-blown panic attack or intense anxiety developing around things that are seemingly minor. For example, you may be driving in your car and have the need to pass a slow driver and the act of switching lanes and passing that slow driver may result in you experiencing intense anxiety. Or maybe you are sitting on your couch and you hear a loud bang from your family in the other room and your heart rate skyrockets and you have a momentary feeling like you need to run and save yourself. These types of restlessness and responses can be extremely overwhelming and problematic and can leave you feeling like you constantly have to be prepared to have an intense response to a minor thing because it feels uncontrollable.

In addition to intense responses, you may also feel constant feelings of worrying thoughts in your mind, even if you cannot logically back up your worry with rational or reasonable reasons as to why. You may feel like you are constantly worried about things, even if you can logically understand why these feelings of worry are invalid or unlikely to ever manifest as being true. Every time you do have a reason to worry, then you might find yourself having an intense stress-related response because you were already in a heightened state of worry in the first place. As a result, it may feel like you are never able to relax and experience rest or calm because you are constantly overwhelmed and stressed out.

Due to your constant state of anxiety and worry, you may also find yourself being highly irritable and moody. It is actually not uncommon for people with anxiety to also experience intense anger, which is essentially a natural "fight" response from the fight or flight mode. You may find that in most cases you want to run away, but in cases where you cannot or where you feel like you are caught by surprise instead, you get angry. You might find yourself yelling, arguing with others, or simply holding onto feelings of intense anger and resentment on an ongoing basis because of the anxiety disorder that you live with.

Other symptoms of anxiety include concentration difficulties and sleep difficulties. As a result of your body constantly producing so much cortisol and adrenaline, it may feel like you are having a hard time paying attention or making up your mind about things. Engaging in basic conversations may be challenging and it may feel like you cannot stay focused on one task for longer than a few minutes at a time. You may also find like your stress leaks over into your sleep and that you cannot sleep soundly each night, making it challenging for you to feel well rested. Alternatively, you may find yourself so stressed out that all you want to do is sleep because it feels as though your time spent sleeping is the only peaceful time that you have. The constant stress may make it feel like you are chronically exhausted, whether you have slept too much or too little, which can leave you feeling uncomfortable and frustrated all the time.

Recognizing Anxiety in Other People

If your loved one is experiencing anxiety, the signs of their anxiety and the symptoms they experience will be fairly simple to identify. Typically, you can tell if someone else is experiencing anxiety based on their mannerisms and the way that they approach the world around them. If someone is experiencing anxiety, typically they will show it in their body language, their verbal language, and their actions, thoughts, and decisions. If your loved one is experiencing anxiety, it is important that you take the time to understand what they are going through and have compassion for them and their experiences, as anxiety can be a challenging thing for anyone to navigate. Having your acceptance and understanding around their experiences are going to be extremely helpful not only in allowing them to manage their own emotional well-being but also in helping you both manage your relationship together.

A big sign that your friend or loved one is living with anxiety is that they seem on edge all the time. Their body may seem jittery or they may fidget a lot when they are in a situation that is making them uncomfortable and it may seem as though their jitters or fidgeting are uncontrollable. It is likely that your friend does not even know that they are doing this, but instead, they do it subconsciously as they try to deal with their growing energy inside of them that is telling them that there is something wrong. They may also seem jumpy or like they scare easy, which can make it seem like they are constantly living in a state of fear. This may or may not be obvious, depending on how high functioning they are alongside their anxiety.

Another common sign that your loved one is living with anxiety is the complaint of aches and pains, especially when it comes to complaints around chest pains. In some situations, these may be indicative of a larger problem, but in people with anxiety, chest pains are extremely common and do not typically have any medical reason behind why they are happening. If your loved one's pulse is racing and they start experiencing chest pains as a result of this, chances are, they are experiencing fairly intense anxious responses and they need to seek professional help if they have not already.

While some people with anxiety have what is known as "high functioning anxiety" and are able to hide their symptoms, others are not able to and will openly voice their fears or concerns around things in their life. Often, their fears or concerns will seem rather grandiose or misplaced, making those around them wonder why they are so afraid of something that is seemingly small. Even if they are high functioning, they will likely voice their fears or concerns except they will likely do it in a way that makes their fears seem smaller than they actually are. Often, they will voice these feelings in passing, try to minimize them with their language, or voice them as if they are a joke so that they can brush it off if anyone asks them if they are serious about how they are feeling.

Another common sign that can easily be seen by yourself or others who are observing your loved one who may be experiencing anxiety is the need to have things done in a very specific way. People who are anxious often find that they need to have some form of control over everything around them as this supports them in feeling like there will be no "surprises" or unexpected turn of events. By making sure that the actions they take and sometimes the actions of those around them are done in a very specific way, the person with anxiety feels like they are in control over their lives. Anytime something does not go according to "plan", they may seem very distraught and overwhelmed, possibly having a large emotional reaction to the experience that may even seem beyond reasonable. If this is happening, there is a very likely chance that your loved one is experiencing anxiety and is not attempting to be controlling or inconsiderate, but instead is simply trying to control themselves and their experience. In other words, this is not their way of trying to make you feel bad, but instead, it is their way of trying to make themselves feel good.

How Anxiety Was Treated in the Past

Unlike depression, the historical treatments of anxiety were much less cruel and gruesome. In the past, anxiety was believed to be a disturbance of ones' inner peace and it was believed that they required therapies like meditation and prayer to relieve themselves of these disturbances. Ancient Greek Stoics believed that anxiety was a disturbance of the emotional mind, which was believed to be in direct opposition of the development to the rational thinking mind. They believed that through stoicism and other mindfulness development practices, people could overcome their anxiety and live a mentally free and sound life. People who were unable to manage their anxiety were believed to be mentally weak and un-evolved, as they were succumbing to the emotions of their emotional mind and struggling to "harness the power of their rational mind."

During Sigmund Freud's time, Freud described anxiety as having been a buildup of sexual energies which would result in people having too much energy pent up in their body. Without releasing that energy, Freud believed that people would suffer from panic attacks and anxiety, which to his understanding was how the energy was manifesting since it was not being expelled through coitus. According to Freud, people who were suffering anxiety needed to engage in more pleasurable and satisfying sexual lives in order to help them relieve their energy to avoid having further implications with their anxiety disorder.

At a later point in history, psychologists started to believe that anxiety was something that we were conditioned to experience by our ancestors based on their own need to fight for survival. Through this understanding, the earliest stages of modern behavioral therapies began to be applied to individuals suffering from anxiety to support them in treating their anxiety. The common belief was that we needed to continually implement behavioral changes in each succeeding generation to ensure that we were passing down positive changes to the following generations.

How Anxiety is Commonly Treated Today

In today's world, anxiety is often treated in the same way that ancient Greek stoics treated anxiety, except that we have a more modernized understanding of anxiety and the psychological functioning of the brain. In certain circumstances were anxiety has become problematic or "clinical," anxiety will also be managed with specific medications that are intended to help calm the brain so that the suffering individual can avoid having further attacks.

When it comes to treating anxiety in the modern world, there is a three-step progressive process that is used to help people overcome their anxious experiences. The first step in this progressive process is self-treatment, the second is counseling and therapy, and the third is medications. By following this three-step progressive protocol, individuals suffering from anxiety can ensure that they are getting proper support and treatment through the least invasive methods possible.

The first step, self-treatment, is often accomplished through the introduction of meditation, mindfulness, breathing techniques, and other stress management techniques. You might also consider creating a personal support network of understanding friends and family to support you, as having that additional level of help during your particularly challenging moments can be powerful. These strategies can be taught to an individual through a therapy session, or they can be learned online or through one of the many books available on the subject such as this one! By implementing powerful stress management techniques on a daily basis, people who are dealing with problematic anxiety can begin to calm down their minds, slow down the production of cortisol and adrenaline, and start relaxing more. As a result of effective stress management techniques, you are able to experience greater freedom from your anxiety and a deeper sense of personal control over your own experiences with the world around you.

The second step if self-treatment is not working is to get involved in counseling or therapy. Talk therapy is the most common form of therapy used by counselors and psychologists to help individuals suffering from anxiety to gain relief from their symptoms. They will also typically walk you through CBT, DBT, and ACT as additional ways to teach yourself to experience greater calmness in your life so that you are no longer suffering from anxiety. While these techniques can be taught by therapists, in the modern world, you can also access the information surrounding CBT, DBT, and ACT through books such as this one so that you can begin incorporating these strategies into your self-treatment.

The final step of the progressive treatment protocol is to talk to a psychiatrist or your doctor if you genuinely feel as though your self-treatment and counseling strategies are not effective enough in helping you gain relief. There are several types of medication that are known to support the treatment of anxiety including anti-depressants which tend to be the most common solution for most individuals who are seeking treatment.

The Power to Influence

Stay mindful: Stay engaged. Don't mentally check out. Keep your focus on your goals and keep your position. Don't let you get distracted or put off by the other person.

Appearing confident: it doesn't necessarily mean that you feel that way. Luckily, it works pretty well to act. Use a voice tone of confidence. You should be relaxed in your physical way. Give the right eye contact. Don't whisper, or stammer on the floor. Act as if you're precious. Showing trust in any given situation is an appeal for judgment. There is a fine line between seeming too arrogant and too apologous.

Negotiate: Be prepared not to get all you want. Be open to alternatives and offer something else to do or some other way to solve the problem. Turn the tables by asking them to suggest alternatives. Ask for your help and be honest about it.

Learning to set limits, saying no, and asking what you want are important skills to build healthy self-esteem and interpersonal effectiveness. Dear Man of DBT divides it into discrete actions that increase the likelihood of a positive outcome.

Are Emotional Healing Vital?

The prerequisite for physical healing is emotional healing. Research is increasingly showing that the mind's condition sets the tone for the body's condition. Negative years and years can make the body weak and sick. Some of our emotional turmoil is inevitable, of course. We all have emotional pain in our lives as a result of traumatic events.

Healing therapies can break the down cycle when we are trapped in pessimistic and discouraging thought cycles. We experience chronic hopelessness and depression if we hang out too long in these circles of unrest and unresolved conflict. This has a negative impact on the physical body in a way that is as real as a cigarette or staying up until 3 a.m. So how do we break the cycle and get emotional healing on the way?

A Two Pronged Approach

Emotional Healing comes in two ways: to recognize what is wrong, and then to shift your focus to happier things. It's important to face the reality of things that aren't right. Ignoring them will not work. But if you're just hanging out, you're not going to solve anything.

In massage and other healing therapies for the physical body, the best approach to healing is often to deal with it directly. This means a massage therapist will move into the area of dysfunction and release the body and restore circulation by dealing directly with the area of pain. Emotions must be treated in the same way directly. It's like getting a fluff and a buff spa massage when you really need deep tissue to deal with them by ignoring them or blurring them in your mind. The first step towards emotional healing is to recognize and face emotional trauma head on. This can occur through counseling and psychotherapy, meditation, journaling, or other ways to focus on the places of pain.

And why we have to move on but we don't have to hang out there. If we focus on life's thorns, we're never going to see the roses. They grow together and it's our choice to concentrate most of the time. Good luck is a choice. This is where friendships are great therapies for healing. Within friendships, mutual giving and taking can do wonders for depression and anxiety, not only by facing it, but above all by moving through it to a better place in your life. The stuff of friendship is joy and happiness. Those who nurture friendships have less anxiety, darkness, and desperation.

Other Tools that enhance your journey to emotional healing In addition to these two basic approaches, the use of plant chemicals such as herbal preparations and essential oils can make our journey to emotional health much easier. Psychotherapy with acupuncture, such as EFT, can be extremely helpful, as can Cranial Electrical Stimulation.

The Journey Now is the time to begin the journey to heal our emotions. Those who take responsibility for their lives come to emotional healing. If we don't, no one is going to do it. So let's do our homework and plan in our lives today a few healing therapies.

Cognitive Behavioral Therapy

The cognitive model of emotional response that is cognitive behavioral therapy is a great one because its basis is that the thoughts of a person are sufficiently powerful to change their behavior, feelings, and point of view. CBT is known as the type of therapy that delivers great results in a shorter time than other therapies and treatments. CBT is also a time-limited therapy, but it is also an endless process. CBT allows a patient to perform their own tasks and therapies when they are at home. During therapy sessions, the results of the task or assignment are discussed and explained.

Apart from regular therapy sessions, there are plenty of useful professional and self-help books on cognitive behavioral therapy that will help a person with emotional Stress. CBT has some techniques and perceptions. Each session has a different plan schedule. It is very important to ask questions in each CBT session. The patient has to ask all of her questions and the therapists have to ask the patient some questions in order to be able to address some things that the patient may not be able to express because there are times when a patient may not be able to tell what she really feels.

Through cognitive behavioral therapy, great approaches are learned such as being able to change the way people think and behave. Positive thinking is also part of CBT, which includes preventing automatic negative thinking and putting an end to it. The other effective strategies and techniques slow down one's style of speaking and walking. A patient must also be relaxed and calmed, as if relaxed and calmed, it is easier for the therapy to sink into the brain of a person. Also, when a person is not stressed or pressured, the information given to patients is processed easier.

There are many websites providing useful cognitive behavioral therapy information. Explanations and examples are also posted on CBT and treatments. Situations and scenarios are provided so that people's family and friends with emotional and behavioral problems get an idea of the problem and how to solve it. CBT history and other studies are also available. Furthermore, articles and blogs can be found on the internet to provide useful information on this therapy. Some websites even offer CBT newsletters and forums for people to interact and share their experiences and thoughts.

The three central ideas of CBT (Cognitive, Behavioral, Therapy)

Cognitive techniques

A key cognitive idea in CBT is 'guided revelation.' This is a therapeutic position which includes attempting to comprehend the patient's perspective on things and help them grow their speculation to end up mindful of their fundamental suppositions and find elective points of view and answers for themselves. A part of the guided revelation is Socratic addressing, which is a strategy for addressing dependent on the manner by which Socrates helped his understudies to arrive at a resolution without straightforwardly letting them know. Specialists use inquiries to test a patient's presumptions, question the reasons and proof for their convictions, feature different points of view, and test suggestions. For instance, 'What else would we be able to accept?' 'What do you think causes ...?', 'What elective methods for seeing this are there?' and 'Why is ... important?'. Guided revelation is integral to the intercessions gone for each degree of comprehension.

Behavioral techniques

Movement planning and reviewed task mean to improve working and deliberately increment pleasurable or beneficial encounters. Movement planning is utilized to design every day ahead of time. The specialist and patient work to lessen the mass of assignments to a reasonable rundown, which expels the requirement for rehashed basic leadership. The evaluated undertaking assignments make reasonable strides to help beat lingering and nervousness inciting circumstances. These methods include getting a gauge of exercises during a day or week, rating exercises on the level of dominance as well as delight, and after that cooperatively structuring changes that will reactivate the patient, animate a more prominent feeling of satisfaction throughout everyday life, or change examples of detachment or stalling. These methods help patients restore day by day schedules, increment pleasurable exercises, and manage issues and troublesome issues by expanding critical thinking.

Therapy techniques

CBT is a talking therapy. It can help individuals who are encountering a wide scope of psychological wellbeing challenges. What individuals think can influence how they feel and how they carry on. This is the premise of CBT. During times of mental trouble, individuals ponder themselves and what befalls them. Considerations can wind up outrageous and unhelpful. This can compound how an individual feel. They may then carry on in a manner that drags out their misery. CBT specialists help every individual distinguish and change their extraordinary reasoning and unhelpful conduct. In doing this, the outcome is regularly a significant improvement in how an individual feels and lives. Intellectual and social psychotherapies are the extent of medications subject to thoughts and principles got from mental models of human inclination and direct. They join a wide extent of treatment approaches for an anger issue, along a continuum from composed individual psychotherapy to self-improvement material.

Why CBT?

CBT is utilized to treat a wide scope of issues. It's frequently the favored sort of psychotherapy since it can rapidly enable you to distinguish and adapt to explicit difficulties. It mostly takes lesser periods than different kinds of treatment and is done in an organized manner.

CBT is a valuable instrument to address enthusiastic difficulties. For instance, it might support you:

- Treat a psychological sickness when drugs aren't a decent choice
- Cope with distress or misfortune
- Manage manifestations of dysfunctional behavior
- Prevent a backslide of psychological sickness side effects
- Identify approaches to oversee feelings
- Resolve relationship clashes and adopt better approaches to impart
- Learn systems for adapting to unpleasant life circumstances
- Overcome enthusiastic injury identified with maltreatment or savagery
- Cope with a medicinal sickness
- Manage ceaseless physical side effects

Emotional wellness issue that may improve with CBT include:

- Anxiety issue
- PTSD
- Phobias
- Depression
- Sleep issue
- Substance use issue
- Obsessive-compulsive disorder (OCD)
- Eating issue
- Bipolar issue
- Sexual issue
- Schizophrenia

Benefits on lifestyle

A feeling of brought down confidence and downgrading of self-influences numerous individuals in our general public. One school of brain science that has been found to help individuals with this issue is CBT. CBT is a type of treatment which spotlights on and has the objective of helping individuals conquer enthusiastic issues.

What describes CBT as a therapeutic approach? "Cognitive" applies to our psychological procedures, for example, memory, discernment, and inspiration. Behavior obviously alludes to our activities and direct. The treatment has been depicted as a careful and deliberate method for moving toward a real existence issue, for the most part, in the region of feelings.

One fundamental quality of CBT is that it depends on logical investigations. Not just have researchers had the option to test and create handy ways to deal with life issues, patients in this type of treatment habitually build up a to some degree logical frame of mind toward themselves. As researchers will test a hypothesis, people may test an observation or memory of an individual or occasion in their life. They may ask, "Is this reality, or am I misshaping my recollections or observations"?

In CBT, the response to this inquiry is critical. On the off chance that somebody encounters contrary feelings toward another, much of the time, that individual is instructed to ask themselves, "what proof do I have that this other individual is attempting to irritate me"? Furthermore, further, "is it conceivable that I may add something extra to the circumstance and ascribing contemplations and intentions to others when, in actuality, these convictions also mirror my very own worries"?

CBT additionally perceives a specific philosophical way to deal with life. CBT addresses convictions that individuals hold about themselves as well as other people. A significant piece of this treatment is to look at the truth of these convictions. CBT is coordinated toward helping individuals be adaptable and self-aiding as opposed to pointless. This type of treatment, likewise, at last, includes conduct. The hypothetical supporting is that intuition prompts sentiments, which at that point, lead to conduct. In the event that somebody's idea examples can be transformed, it pursues then that their sentiments and conduct can be changed moreover. The objective of the treatment is to realize changes in idea designs so that emotions become increasingly positive, and conduct winds up valuable toward oneself as opposed to destructive as when negative sentiments are followed upon.

CBT stresses an engaged methodology. An individual using CBT treatment is urged to move from reckless passionate and conduct issues toward the heading of defining objectives, including all the more fulfilling sentiments and conduct. Subsequently, CBT is a deliberate, critical thinking way to deal with the enthusiastic issues that go with life for us all.

Numerous individuals accept that they feel a specific way, for instance, furious, in view of something that another person has done to them. After some investigation, they can see that they have credited significance to what the other individual has done, and they have enabled themselves to lose control in response to the next individual. They could have likewise picked some other method to respond. An elective response might be more beneficial and include imagining that produces sentiments and conduct that are substantially more fulfilling than earlier reckless practices.

Truth be told, foolish conduct has much of the time been the reason for the inception of CBT. CBT has been appeared to create very great outcomes with evacuating pointless practices and is consequently experiencing expanding acknowledgment in the mental network.

Benefits on health

Highlighted below are some key ways CBT benefits patients from various backgrounds:

Brings down Symptoms of Depression

CBT is a standout amongst other known, exactly upheld medicines for depression. Studies demonstrate that CBT enables patients to conquer manifestations of misery like sadness, outrage, and low inspiration, and brings down their hazard for backslides later on.

CBT is accepted to work so well for easing sadness since it produces changes in perception (considerations) that fuel horrendous cycles of negative emotions and rumination. CBT is so defensive against intense scenes of wretchedness that it very well may be utilized alongside, or instead of, upper prescriptions. CBT has additionally demonstrated guarantee as a methodology for helping handle postpartum depression and as a subordinate to prescription treatment for bipolar patients.

2. Diminishes Anxiety

There's solid proof with respect to CBT treatment for anxiety-related issues, including frenzy issues, summed up uneasiness issues, social nervousness issues, fanatical impulsive issues, and post-horrible pressure issues. Generally speaking, CBT exhibits both viabilities in randomized controlled preliminaries and adequacy in naturalistic settings between patients with tension and advisors.

Scientists have discovered that CBT functions admirably as a characteristic solution for anxiety since it incorporates different blends of the accompanying strategies: psycho-training about the idea of dread and tension, self-observing of side effects, substantial activities, psychological rebuilding (for instance disconfirmation), picture and in vivo presentation to dreaded improvements (introduction treatment), weaning from inadequate security flag, and backslide anticipation.

3. Helps Treat Eating Disorders

Dietary issues give probably the most grounded sign for CBT. It has been found that CBT can address the hidden psychopathology of dietary problems and question the over-assessment of shape and weight. It can likewise meddle with the upkeep of unfortunate body loads, improve drive control to help stop voraciously consuming food or cleansing, decrease sentiments of disconnection, and help patients become increasingly agreeable around "trigger nourishments" or circumstances utilizing introduction treatment.

Cognitive therapy has turned into the treatment of decision for treating bulimia nervosa "eating disorders not otherwise specified" (EDNOS), the two most basic dietary problem analyze. There's additionally proof that it tends to be useful in treating around 60 percent of patients with anorexia, viewed as one of the hardest psychological maladjustments to treat and keep from returning.

4. Diminishes Addictive Behaviors and Substance Abuse

Research has demonstrated that CBT is powerful for helping treat cannabis and other medication conditions, for example, narcotic and liquor reliance, in addition to helping individuals quit smoking cigarettes and betting. CBT is exceptionally compelling in decreasing backslides in nicotine slackers and appears to be better than other helpful methodologies. There's likewise more grounded help for CBT's social approaches (halting driving forces) in the treatment of tricky betting addictions contrasted with control medications.

5. Develops Self-Esteem and Confidence

Regardless of whether you don't experience the ill effects of any genuine mental issues whatsoever, CBT can enable you to supplant dangerous, negative contemplations that lead to low confidence with positive certifications and desires. This can help open better approaches to deal with pressure, improve connections, and increment inspiration to attempt new things. The Psychology Tools site gives incredible assets to utilizing CBT worksheets all alone to take a shot at creating positive relational abilities, sound connections, and supportive pressure lessening strategies.

Why is CBT so popular?

CBT has been getting increasingly more consideration as of late. In the event that you've found out about CBT and are thinking about what all the buzz is about, here is a synopsis of why CBT is ending up progressively prevalent.

A New Face for Psychology

For a long time, there has been a mainstream picture of individuals going to see a clinician for quite a long time so as to attempt to get to the foundation of every one of their issues. While there may, in any case, be a few therapists who have similar individuals visiting them and lying on their love seat for quite a long time, those are surely the exemptions.

In the event that you ask a top therapist, the person will clarify the fame of CBT is because of its prosperity. It offers snappier and all the more verifiably positive outcomes. As opposed to inactively lying on a lounge chair, an individual getting CBT effectively works with the specialist to create arrangements. CBT has a demonstrated record of progress, and it tends to be utilized to address a wide scope of concerns, including rest issues, fears, dietary issues, and sadness.

Availability

A central point behind the notoriety of CBT is its availability, as CBT is all the more promptly accessible to individuals looking for treatment. Furthermore, CBT can create generally quick outcomes, which means it might be more moderate than increasingly customary time-escalated treatments. All the more along these lines, courses of treatment following the CBT model can be time-restricted and outfit individuals with abilities inside a couple of months. Contingent upon the issue being worked through, a course of CBT could last somewhere in the range of six and twenty sessions.

Furthermore, to utilize alternate importance of the word availability, the standards of CBT are more effectively comprehended than increasingly complex hypotheses and are hence generally simple to learn. This enables the client who is attempted the treatment.

Skills-based approach

Connected to this thought of strengthening is the way that CBT is abilities based and adopts an instructive strategy. It comprises of being acquainted with systems that can be utilized to handle issues, for example, negative reasoning, and working on utilizing them, with the help and help of the specialist. It might include "schoolwork" of activities to do at home, for example, making a log of sentiments. The emphasis is on a client learning aptitude required for them to handle their very own issues, and tackle issues individually. These systems, once learned, can be utilized later on and are valuable when different issues come up. This implies the advisors' job is to help outfit clients with abilities and help them practice these, instead of being a specialist that "fixes" issue. Along these lines, a few people will discover this methodology less undermining, and furthermore decreases the danger of passionate reliance upon an advisor.

Research proof

Another significant factor is CBTs ongoing phenomenal reputation of research demonstrating its viability. In the last official statement, it was referenced that the institutionalized estimates utilized in CBT implied it was increasingly quantifiable, and value could be estimated all the more effectively. CBT is useful in treating a significant burdensome issue, social fears, summed up tension issue, alarm issue, post-horrendous pressure issue, bulimia, and a portion of the conduct issues that mentally unbalanced kid involvement. There is additionally some proof to propose is can be helpful in treating a bipolar issue, over the top impulsive issue, anorexia nervosa, cocaine misuse, and sexual issues.

Quantifying improvements

CBT looks to gauge the additions clients profit by in noticeable and solid terms. For instance, during CBT, clients might be asked to every now and again rate the force and events of negative contemplations. An individual encountering uneasiness might be solicited to measure the power from the on-edge emotions from 1 to 10 while pondering a circumstance that makes them anxious. In the event that, during or after a course of CBT, they rehash the activity and rate their degree of negative emotions lower, this is a sensibly quantifiable, yet abstract, show that the client feels they have experienced positive change. This implies it is conceivable to see upgrades made through CBT in quantifiable and more logical terms than it is in different types of treatment.

New Awareness

Before, issues like depression and rest aggravations were either disregarded or treated with the drug. Throughout the years, the dangers and impediments of medicine have turned out to be better comprehended. While medicine can be an extraordinary choice for specific conditions or for a constrained measure of time, individuals have turned out to be progressively persuaded by increasingly all-encompassing, long haul draws near. Presently a developing number of individuals are holding onto CBT as an approach to improve their everyday lives, musings, sentiments, and practices without an over-dependence taking drugs.

Individuals are additionally winding up progressively mindful that CBT can be valuable when utilized nearby restorative treatment. People managing interminable medical problems or with the changing strength of a friend or family member are presently observing the advantages that CBT can give. Improving methods for dealing with stress and discovering progressively positive approaches to manage pressure and nervousness can give gigantic functional assistance.

Organizing Health

Much the same as individuals go to see their medicinal specialist when they have an area of concern, and more individuals are assuming responsibility for their psychological wellbeing by looking for expert assistance when required. Before, numerous individuals felt that solitary individuals with especially major issues expected to see an analyst. Presently they understand that there is no motivation to battle alone with melancholy, fear, or different issues. CBT can enable them to determine these issues and have a more joyful existence.

Does it work for me?

CBT has been utilized to help patients who are experiencing depression, tensions, addictions, and a wide range of other psycho-social issues. When encountering CBT, specialist motivates the suffering individual to address It is recognized that cognitive models and the way by which an individual may identify with express situations are associated with the patient's feelings and conduct.

CBT is a way to deal with assistance, find the basic explanations behind the issue from a psychological point of view, and thereafter change or right the thinking model that has incited wrong lead. Utilizing CBT, an expert is attempting to alter the ridiculous and misshaped thinking about the patient. This thusly will support the patient with making changes in conduct and have the choice to re-alter. Thinking examples and feelings undertake a crucial job in human conduct and can be adjusted or changed.

CBT is as well known to assist someone with chronic drug habits like heroine. In the severest feeling of the word drug addict, both physicians recommended drugs that are compelling, just as illicit drugs, can be said to have a conduct issue and can profit by CBT. There is an increasing number of individuals undergoing broken issues, and keeping in mind that some accept restorative medications might be sufficient. Studies appear to show that intellectual, behavioral therapy is fruitful. Obviously, a great deal relies upon the individual's eagerness to follow a prepared advisor and to adjust internal considerations and emotions.

The prepared advisor additionally helping the patient to fathom past conditions and experiences. CBT has transformed into a technique for understanding the relationship between internal thoughts and acknowledgments and human lead. This no susceptibility has added to some accomplishment that has been made. It is like manner has helped a couple of individuals to turn out enormous enhancements for a mind-blowing duration.

In case you are a person who is encountering apprehension or distress or some other kind of psycho-social issue, take bravery and find a readied counsel in CBT. You can make sense of how to make changes for an amazing duration and help yourself and the people who are close to you. Clearly, it may put aside you some push to see a differentiation in your life, anyway make a point to accomplish anything useful you need affirmation.

The time you may spend can have an impact. Luckily, paying little respect to whether you feel overwhelmed and demoralized now and again, there is help for you. There is, in like manner, help in kind of workshops that you can deal with becoming acquainted with CBT and how it can strengthen you. Putting aside some push to observe the information obtained may be your supreme starting advance to recovery. Thus, in any case, your circumstance, CBT worked for me and unquestionably getting down to business for you also.

Dialectical Behavioral Therapy Strategies for You to Implement Today

The following practices are DBT strategies that you can begin implementing in your life today so that you can start experiencing the positive benefits of this particular therapy practice right away. Remember, DBT does rely on you triggering your unwanted emotional responses to a degree, so it is important that you are able to do so in a way that is not going to result in an overwhelming or uncontrollable emotional response. It is truly imperative that you proceed in the proper step-by-step method if you are going to generate success through DBT practices. The slower and more intentional you are at implementing these practices, the easier it is going to be for you to generate success using this modality.

Step 1: Mindfulness

The first step in DBT, as you know, is mindfulness. During the mindfulness phase, your goal is to gain a complete sense of self-awareness so that you can start identifying your ineffective behaviors and recognizing how they are negatively impacting your life. When you can identify these behaviors in yourself, you can also identify how you may be worsening them and what you may be doing that is blocking yourself from effectively working through these emotional experiences.

The best way for you to begin developing mindfulness is through meditation and regular self-auditing so that you can start developing a clearer sense of who you are and what your energies feel like. If meditating is particularly challenging for you, consider following a guided meditation on YouTube so that you can experience actual guidance through the process. For many people who are just starting out, having someone actively walking them through the process step-by-step can be helpful as it ensures that you have something to focus on. You may benefit from using the same meditation over and over for a while until you can effectively follow the meditation the entire way through without becoming distracted or ending it early due to mental or emotional discomfort. This way, you can start training your mind to relax to the same listening experience, which may make it more comfortable and tolerable for you over time.

Another popular practice that is used in developing mindfulness with DBT is the raisin exercise, which you may have already heard of. In the raisin exercise, your goal is to slow down the process of eating a raisin so that you can mindfully engage in every single step of the process. Your first step in the raisin exercise is to look at the raisin and really become intentional over identifying what the raisin looks like. Mentally describe its appearance to yourself as you observe it, creating a full mental description of the raisin in your mind. Then, pick up the raisin and engage completely in the process of touching it and feeling what the raisin feels like. Next, smell the raisin and give yourself a moment to identify what its aroma is like or if you can even smell one. If you can, see if it invokes any memories or feelings within you that give you a certain feeling inside. Then go ahead and put the raisin in your mouth, but do not chew it just yet. Instead, move it around in your mouth and start feeling what it feels like before you chew it. Lastly, slowly start chewing it and be very mindful over what it feels like with every single bite so that you can start getting a deeper sense of what it tastes like and feels like. When you are done chewing, swallow the raisin and spend some time reflecting on what your experience was like and how it felt for you.

Using music is also another highly effective way to start building up your mindfulness practice so that you can begin identifying your inner thoughts and feelings more effectively. With mindful listening, your objective is to put on a song that you have never heard before and begin listening to it without any intention of judging it or mentally responding to it whatsoever. As you listen to the song, simply take it in, listen to the words, and see if you can identify what instruments are being used to create the beat to the song. Do not determine whether or not you like the song, what you think or feel about the singer's voice or anything else of that nature. Instead, simply listen to the music and seek to experience it for what it is, without trying to manipulate your experience in any way through your own thoughts.

Step 2: Interpersonal Effectiveness

The next step in DBT is building up your interpersonal effectiveness or your ability to engage in positive and healthy relationships with other people in your life. For interpersonal effectiveness, you want to be focused on having more positive relationships with your friends, family, co-workers, boss, and anyone else with whom you might share a relationship. When it comes to nurturing these relationships, your goal is to stop engaging in ineffective behaviors that nurture the negative aspects of these relationships and to start engaging in positive ones that nurture the positive aspects of these relationships.

The interpersonal effectiveness stage of DBT can take a while, as the relationships that you share with others require two separate people to be contributing to them in order for them to exist. That being said, your relationships are not exclusively with yourself so you need to spend a significant amount of time here taking responsibility for your own engagements in relationships and learning how to disengage from unhealthy behaviors. For example, if you are experiencing an argument with your sibling, you are not able to control your sibling so your goal, instead, would be to take responsibility for your part in it and learn how to assert your boundaries effectively.

When it comes to developing better interpersonal skills, your best skill to work on is going to be your boundaries. This will come in two levels: your boundaries in how you are willing to let others treat you and your boundaries in how you are willing to treat others. In being able to moderate how you are allowing others to treat you and vice versa, you create a more powerful opportunity for you to have impactful, meaningful, and effective communications and relationships with those around you. This is because, more often than not, intense emotional responses to triggers in relationships are related to feeling as though your personal boundaries are being breached, which can result in feelings of anger and distress. For people who are not used to having effective boundaries in place protecting the best interest of both parties in the relationship, it can be easy to jump to feelings of anxiety or depression because you do not feel safe in the relationship. Instead, you feel like everything triggers you to feel vulnerable or exposed and you struggle to trust yourself and your own ability to protect yourself and your needs as a result of these vulnerable experiences. As a result, you end up having intense overreactions to the stimuli because you are not able to regulate your response and choose one that is more appropriate to the situation. In your mind, it is an uncomfortable and dangerous situation and it feels far more intense than it really is.

This is where boundaries come into play. When you have boundaries and you learn how to assert these boundaries effectively, you begin to trust in your own ability to navigate relationships effectively because you are able to trust in your coping methods. Through this, your reactivity in relationships begins to slow down because you are more confident in yourself and your ability to avoid painful unwanted experiences. So, in order to create this sense of comfort, you need to discover how you can create and assert boundaries in your relationships.

Creating boundaries in relationships takes a lot of willingness to deeply align with your inner truth and speak that honesty to the people with whom you are sharing a relationship with. In developing a stronger ability to communicate, asserting your boundaries becomes easier because you feel more confident in your ability to understand your own needs and communicate those to other people. Early on, you may need to practice this in calmer or less stressful relationships to ensure that you are not trying to put too much pressure on your more intense relationships. For example, if you have a more casual relationship with your friend but a highly reactive one with your Mom, start practicing in your relationship with your friend first. This way, you can begin to develop confidence in yourself and your own ability to assert yourself first, so that when you go to assert yourself in the relationship that you share with your Mom, you have a greater sense of confidence in yourself. This will ensure that you feel capable of speaking up and protecting your boundaries, even if it feels hard.

Aside from communication, you also need to ensure that you are actively following through with what you say you will do. Do not assert your boundaries only to fall weak on them once you have declared what they are. If you have asserted that you will not tolerate someone belittling you or abusing you, then you need to ensure that you no longer tolerate that in your life. If someone in your life does start belittling you and refuses to listen to you, you need to be prepared to walk away and disengage from that relationship with the said person. Knowing how to walk away is imperative, as this is your follow through. It is also important that you realize that early on, there will be many instances where you need to assert yourself more intensively than you will later on in your life. At first, the people in your life are going to take time to adapt to the new reality, that is you no longer allowing people to treat you the way that they have been treating you. These people are not used to treating you in this new way, so it will take intentional and intense follow through on your end to ensure that they get the message and start changing the way that they treat you unless they want the consequences. Make sure, however, that the consequences are peaceful and not reactive, as this will only further engage in the behaviors that they are used to. Over time, people will begin to understand your changing of standards for how you are willing to be treated and will begin to realize that your boundaries are a certain, non-negotiable standard in your life. As a result, you will not have to work so hard to assert yourself and dole out as many consequences to people who are not respecting you and your boundaries.

Step 3: Distress Tolerance

Distress tolerance means to develop a tolerance towards being in stressful situations that are known to trigger your unwanted emotional experiences like anxiety and depression. When you are dealing with anxiety and depression, building up your distress tolerance means building up the skills that you need to overcome the triggers that are known for triggering your anxious thoughts or depressive episodes. Building your distress tolerance can be stressful because you will need to intentionally expose yourself to these distressing situations in order to develop your tolerance. Exposure is one of the primary things required in building your tolerance, alongside developing effective tools to actually manage and navigate your triggers effectively. So, for that reason, it is important that you take this process very slowly to avoid overwhelming yourself and re-traumatizing yourself around your known triggers. In going slowly and intentionally, you can ensure that you are at less risk of having intense reactions because you remain in the feeling of being in control over yourself during your exposure. Although your feelings of anxiety or depression will begin to stir, you will not stir them so much that you lose control, instead, you will stir them until you are just about to lose control and then you will comfort yourself from that situation. As a result, your tolerance begins to build up because your trust in your ability to comfort yourself and stay safe and protected in these triggering situations grows as well.

Building up your tolerance requires you to invoke as many of your senses as possible into the calming experience so that you can feel completely in control over yourself and your body. So, each time you enter a situation where your anxiety or depression is being triggered, as soon as you recognize it, start taking intentional control over yourself and your reactions. Focus on pacing your breathing, relaxing your muscles, and adjusting your physical sensations so that you can experience a greater sense of calm. If you can, change the temperature of your environment, rub your hand against something soft, or even start chewing a piece of gum so that you have something new to focus on. By adjusting your physical feelings and sensations, you can help your mind begin to gradually shift focus and start approaching previously triggering situations with a greater sense of intention and calmness.

As you continue to expose yourself to these distressing situations and intentionally calming yourself down, you will discover that you can go into situations that are more and more intense. This is proof that your practice is working. That being said, you may also find that sometimes a trigger that you could previously tolerate is no longer tolerable as you begin to feel overwhelmed by the environment that you have now entered. This is completely normal, especially for people who have had intense and prolonged reactions to the world around them. If you have had anxiety for a long time, for example, relieving yourself from your anxiety completely is going to take time and there will occasionally be triggers that feel bigger than what you can reasonably handle. Simply keep working on intentionally building up your ability to soothe yourself in these challenging situations, exercising your interpersonal effectiveness, and behaving in a mindful manner and you will find that each situation gets easier and easier. In no time at all, situations that were previously highly triggering for you will begin to feel calmer and easier to manage.

Step 4: Emotional Regulation

Emotional regulation is the final step in building up your DBT skills and it is necessary for helping you experience the emotions associated with your reactivity without experiencing them in such an intense manner. For example, if you live with anxiety, you may find that every single time you experience even a small amount of anxiety, it quickly escalates because you are habitually used to letting it escalate. This is a normal experience in people who experience reactivity or intense emotional responses to certain triggers because they become so used to responding in such an intense manner that regulating their emotions becomes challenging. You may also find that all secondary emotions relating to your initial response are also more challenging for you to navigate since they are being fueled by the original more challenging emotion.

Emotional regulation can be built up in many different ways, though one of the most powerful ways continues to be mindfulness, meaning that the very foundation of your DBT practice is already fueling your ability to emotionally regulate at this point. In addition to your mindfulness practice, you can begin including things such as mantras, exercising, and journaling. Each of these practices is powerful for their own unique purposes and they can all easily be used together to help you achieve a deeper sense of emotional regulation.

Mantras are powerful as they can be used during periods where you are feeling your emotional reactivity being triggered and you need to have the ability to stay in control over yourself. Rather than repeating something like "I can't handle this" or "I'm too weak to change," you can repeat something like "I choose peace." These types of mantras can be repeated whenever you feel as though you are struggling to stay in control over yourself so that you can hold onto your original goal and stay in alignment with your key desires.

Exercising is powerful as emotions are known to develop intense surges of energy in the body, which needs to be used in one way or another. In your ineffective emotional regulation patterns, this energy was likely being used to engage in your intense emotional reactions to the environment around you. In a more effective situation, you could choose to engage in the situation with a greater sense of peace and then exercise to eliminate the built up energy that you did not need to use. So if you experienced anxiety, you could choose to navigate the situation causing anxiety with peace and then go exercise to release the shakiness and discomfort that built up within you from your anxious response.

Journaling is an exceptional way to develop self-awareness, a deeper understanding of your own emotional wellbeing, and the ability to recognize where your emotions come from and release them completely. In a lot of circumstances, intense emotional responses can be triggered by not feeling as though you are understood or heard by those around you, so by journaling you can get it off of your chest and offer yourself the understanding that you need. This may not be the understanding or compassion that you were seeking, but in time, you will find that it is a lot easier (and more reliable) to seek your own compassion and understanding than it is to seek it from those around you. Furthermore, this consistent feeling of being understood, released, and cared for will result in you not feeling so overwhelmed in emotional relationships with other people.

Who Can Benefit from DBT?

DBT is most effective for people who experience emotions very intensely. They tend to be easily overwhelmed by life and relational stressors to the point that they feel that their emotional responses are out of control. Consequently, they often act in an impulsive manner in an attempt to temporarily relieve some of the distress. However, their reactions over the long haul tend to create additional problems.

DBT was initially used to help individuals who were diagnosed with Borderline Personality Disorder (BPD.) It was and continues to be a very effective therapy to use with this population. However, in more recent years, DBT has also been very successfully used with other individuals who demonstrate severe mood swings and are unable to apply coping strategies to successfully deal with these intense and sudden emotional urges. Many of these people are struggling with severe depression, PTSD, eating disorders, severe compulsory disorders, Bipolar Disorder, ADHD, anger management, and substance abuse. Many people who seek out DBT also engage in self-harm as this therapy has been shown to be very effective in helping individuals with this level of emotional trouble.

In order to really understand more about the person who generally does well in DBT, let's take a look at the specific characteristics that many of these people share. People who do well with DBT usually have a high level of *emotional vulnerability.* What this means is that they are prone to experiencing emotions in a very reactive and intense way. Sometimes they are just hardwired to feel emotions more intensely than the average person. In fact, DBT theory asserts that the Automatic Nervous System of an emotionally vulnerable person is predisposed to be reactive to relatively low levels of stress. Their nervous system also takes much longer to return to baseline when the stressor is removed than other people. Additionally, some people have mood disorders such as Major Depression or Generalized Anxiety that is not being effectively controlled by medication that influence how intensely they experience their emotions. Consequently, emotionally vulnerable people tend to have quick, intense, emotional reactions that are difficult to control. This keeps them on a roller coaster ride throughout life.

However, clinicians have found that most emotionally vulnerable people who seek DBT treatment are not JUST hardwired to have more intense emotions or have mood disorders. Typically they have also been exposed to environments that were not validating for extensive periods of time, usually beginning in early childhood but can occur at any point. These environments did not provide them with the support, attention, respect, or understanding that they needed to properly work through their emotions. Invalidating environments can range from severe emotional or physical abuse to mismatched parent and child personalities. Consider the shy child who is born or adopted into a family full of extroverts who is constantly teased about his introverted personality. Or perhaps it's the child with ADHD who has a mother and stepfather who are inflexible and constantly yelling at her. These are both examples of environments that are invalidating. When the person who already has the predisposition to experience more extreme emotions is placed in an environment that does not support or validate his feelings, he can become even more emotionally vulnerable. He may then begin to demonstrate even greater emotional reactivity because he inadvertently learns that the only time that he is taken seriously is when he demonstrates this extremely emotional behavior.

Let's take the introverted child as an example. Say that he was constantly told by his father that he had to "man up" and become more aggressive in his approach to life. He felt ridiculed and began to think that there was something wrong with him. So one day when his father was addressing him, the boy began to cry uncontrollably. His father immediately lightened up and his mother came rushing to his aide, showering him with tons of attention. Then it happened again. And then again. An interesting thing began to happen. The young boy's unconscious mind began to see a pattern that looked like this: My dad harasses me, I cry uncontrollably, the hassling stops, I get lots of attention. He begins to do it more and more because it works and each successful demonstration inadvertently reinforces the behavior. His emotional outbursts have become validated and then it ultimately becomes an ingrained coping skill.

The process described above unconsciously reinforced the child's emotional vulnerability and you guessed it - made it worse. This is typically the pattern that you see with people with Borderline Personality Disorder, Bipolar Disorder, eating disorders, and other disorders that DBT treats. The next section will give an overview of several of the disorders that are successfully treated with DBT Therapies.

Borderline Personality Disorder

People with Borderline Personality Disorder (BPD) experience emotions more intensely and for longer periods of time than other people. They are prone to frequent and chronic outburst to the point that many Mental Health Professionals have described this population as one that experiences unrelenting crisis. They are nearly always in crisis mode as they have generally not learned the coping skills that they need to better regulate their intense emotions.

Individuals with BPD are emotionally vulnerable and it takes them much longer to return to baseline following an event. In addition, one of the patterns that Therapists have described regarding people with BPD is their tendency to take on the same belief system as the invalidating environment that they are subjected to. This results in their own "self-invalidation" where they reject their own emotions and ability to solve problems. They also tend to develop unrealistic expectations for themselves and experience intense shame and anger when they fail to meet their goals or when difficulties arise.

Another defining characteristic of individuals with BPD is their tendency to make rigid and unrealistic demands of themselves and others. When things don't go as they plan or desire, then they often resort to "blaming." Blaming is a thinking error that many people with BPD have. They blame everyone and everything for their problems and have trouble acknowledging the personal behavioral changes that need to be made to see different outcomes in their lifestyle.

Individuals with BPD have a poor sense of self and tend to struggle with interpersonal relationships. They tend to seek out individuals who will take control and solve their problems for them so that they can shrink back and not have to do so. However, they tend to wear the mask of competence so that others think that they are capable of solving their own problems and dealing with their intense emotions. Although they have mastered certain areas of their life, they have not been successful at generalizing their competence in other areas of their life.

Due to the lifestyle that most people with BDP have created in combination with their difficulties returning to baseline following an emotional event, they end up experiencing significant traumatic experiences on an ongoing basis. They also tend to avoid experiencing negative emotions all together because they do not know how to regulate even healthy negative emotions. As a result, they do not know what to do when an emotional situation arises that they are unable to tolerate which throws them into an intense and prolonged emotional state.

People with Borderline Personality Disorder sometimes engage in cutting or other self-harm and suicidal behavior to deal with the intense emotional pain. Emotional vulnerability is often seen with individuals who are suicidal or engage in chronic self-injury. The individual is highly emotionally reactive and when exposed to very severe trauma such as physical or emotional abuse, she begins to think about suicide. Eventually, in an attempt to be relieved of the ongoing pain, she tries to kill herself and is taken to the hospital. Here she is given loads of attention and for the first time she begins to feel like she is being validated and taken seriously.

Then consider the boy who engages in some type of self-injury such as cutting or burning himself because he finds that it provides temporary relief. When someone else finds out that this happening, suddenly people are taking him seriously. Similar to the experience of the girl in the first example, for the first time he is finally feeling validated.

What do you think happens in these two situations? Over time they both continue to engage in the behaviors because it is the only time that they feel validated and supported. It becomes an ingrained coping skill.

Building Willpower to Change

One thing that can be rather difficult when trying to make a positive change for yourself is actually taking the first step forward. Once you do take the first step and start to establish some sort of routine for yourself, you settle into it mentally and it becomes a lot easier. It also tends to become even easier as you press forward with it. However, when you're just starting out, it can be difficult to really cement that will within yourself.

Understand that the conduit to real and actionable change springs forth from within yourself. In fact, when working with people through the dialectical behavior method, therapists recognize that the first step is to make sure that the patient is agreeable and wants to get better.

So, with all of that in mind, how can you take the first steps towards recognizing that something needs to change? The first is through an honest analysis of your actions and your behaviors.

The first thing you need to realize is that if there is constant tension in your life, there is a chance that you are a common denominator. There is nothing wrong with realizing this, though. We all carry a large amount of baggage throughout our lives and realizing how that baggage continues to affect you is a rather large part of the healing process.

Recognize that there is nothing wrong with wanting to change for the better. Healthiness is a good thing. You should be actively seeking this healthiness. Although it may not seem like it right now, the type of happiness that you're wanting is both real and attainable for everybody, even yourself.

The first step to getting there, though, is to admit to yourself that you need to change. If you recognize that things are going poorly, you need to analyze the things that are happening to you and honestly try to see what part you're playing in them.

Many people feel like the second that they admit to themselves that they could make a change, their entire life will be thrown out of whack. Sometimes they feel like their current life is too cemented to really make an effective change possible. Both of these viewpoints are explicitly wrong, though.

The truth is that while some major shifts might be necessary in your life in order for you to become a happier and healthier human, the vast majority of these are going to be coming from within yourself - they are not, necessarily, things which occur around you.

I suppose that the best part about it all is that most of the change comes from a shift in your general perspective. The weight that is lifted when you start to move away from those things which fog your general perception, in such a way that you can start to think and act rationally, is a very beautiful thing; you'll feel like a new person.

The biggest part of all of this comes with truly desiring to make a change. Understand that having that true desire for change means necessarily that you are able to accept things about yourself, even the hard things, that are generally difficult for you to accept.

These things may be good or bad, too; for example, if you suffer from depression, it may be that you need to accept that some of the irrational thoughts and conceptions aren't necessarily true. On the other hand, if you suffer from a bit of situational narcissism and tend to displace blame as a defense mechanism, you may find that you need to accept that you can, in fact, do wrong - even if your narcissism is a defense mechanism within itself.

The hardest part of all of this is the fact that mental illness really, ultimately, is not cut and dry. Working with mental illness can prove very difficult in fact, because everybody's mental illness will manifest in a way which is essentially different, even if the illnesses follow similar paths and share certain attributes. Therefore, it can be quite difficult to really get out of these rhythms and figure out what is true and isn't, as well as what will work for you and what won't.

Be aware, though, that all of the change you need will ultimately be summoned from within yourself. You must summon the desire to make a real and genuine effort to change.

You can and most likely will experience success using this method, but you have to be open to change. The big kicker is self-discipline. It's going to take a lot of self-discipline for you to navigate this form of therapy like you're going to need to.

When you suffer from the problems that dialectical behavior therapy is intended to fix, training this amount of self-discipline can be rather difficult in and of itself. It is not, however, impossible. Your mind is constantly frayed or jumping all over the place, or sometimes you have rapid and unbearable mood swings. Even if not those, it's possible that you react to situations in self-destructive ways that you don't quite understand. Perhaps you don't even know what you do or why, but you know that it's having a negative effect on your relationships.

The point of dialectical behavior therapy is ultimately to help you pinpoint, recognize, and then effect change in regard to these problems. In order to do so, you also have to be willing to admit to yourself that these problems exist in the first place. This is the crucial first step to progress. Understand this before you take any further action; reaching the end of the road to success requires that you drive it properly. Driving it properly here means that you need to be prepared to accept that you are not perfect; you will make mistakes, you will have inappropriate or hurtful thought patterns, you will do things that you ought not do. Accepting that you'll do so is absolutely pivotal in the grand scheme of sorting out your mental issues.

Another critical thing is avoiding getting discouraged. Remember that this program specifically lasts for *six months* and is meant to be cyclical such that it can be repeated as necessary. Your results are not going to be instantaneous, and chances are that you're going to regress sometimes too. Your method for dealing with this needs to revolve around accepting the fact that not everything is going to move in a linear fashion.

Realize that you're going to mess up from time to time, and also that you're ultimately not going to make a heap of progress all at once. Consider it like losing weight. When you lose a lot of weight, it can be difficult to really ascertain how much weight you've *lost*, because you see yourself every day. The only time that such a thing really becomes apparent is if you compare yourself to an old picture or if somebody brings up the fact that you've lost weight.

It's much the same here. While a lot of the mental changes won't be immediately noticeable, the changes you make to your mindset as a whole will lead to the development of a far healthier general thought process over the long run.

Practicing Positive Mindfulness

There are four modules that you follow in dialectical behavioral therapy. They can be used to treat borderline personality disorder. In the following, we will look at more depth at the DBT treatment modules for borderline personality disorder, along with examples to guide each one of the modules. We hope that you will find it to be helpful in your continued exploration of dialectical behavioral therapy. We will show what you can do to deal with your complex emotions with the different DBT strategies that are taught by therapists. The first is practicing mindfulness.

Mindfulness

Mindfulness provides a foundation for DBT. It is based on Eastern Zen Buddhism and also includes Western contemplative practices. 25 Arnold T 2008 Core mindfulness _______ Mindfulness enables a person to be aware of their thoughts, feelings, and behaviors. Through this practice, an individual can take responsibility for his or her life in a different way. After all, mindfulness has been proven to be able to help a person regulate his or her emotions and work on accepting feelings and making changes in his or her life.

When doing dialectical behavior therapy, a person will practice mindfulness and be concerned with three different states of mind: wise mind, logical mind, and emotional mind. The first is the ideal state that a person may want to reach in this lifetime. The other two comprise what a person strives for with a wise mind. The logical mind, to be specific, is used when someone does math problems, looks at a map, or performs other rational tasks.

The logical mind deals with empirical facts and concrete tasks that a person can accomplish. The last state of mind is the emotional state, which allows a person to experience the most profound feelings and act out of emotional reaction to a situation. The latter is usually connected to impulsive decisions and is also known as the "hot" state of mind because of the intensity of reaction that someone gives during a heated moment.

In the wise mind, a person is grounded and is aware of emotions and knows how to react to the various circumstances. Even when prompted negatively, he or she can recognize the feelings of anger or frustration and respond appropriately to anyone.

It is essential that the individual learns to be aware of his or feelings. The skills that BPD patients acquire from DBT sessions include observation, description, and participation. 26 Grohol J M 2018 An overvi ew of d Let's look at a more detailed description of these skills.

Observation

The first ability to acquire in mindfulness is observation. In this part, the individual should become aware of thoughts, emotions, situations, and behaviors without changing them. Instead of trying to modify or respond to it, the person directly collects data from experience. This skill includes note-taking to record what is happening around him or her.

Description

After having observed a person's situation, he or she can then describe an occurrence. With more description of the event, an individual can have more empathy and exercise self-control. During this phase, he or she analyzes the situation with the data collected and separates fact from perception.

Participation

With the third skill, the person participates in the current moment and becomes more aware of the present and lives in every moment constructively.

Other skills in which you may use mindfulness include how you judge the world. For example, taking a non-evaluative approach to a tensed situation makes it better and more uplifting for someone. Rather than seeing a situation as good or bad, he or she can holistically look at a circumstance and understand how to deal with it.

Also, a person who practices mindfulness can live in the moment and focus on one task at a time while not worrying about how a situation will turn out. Instead of thinking about many things at once, the individual can do mono-tasking, which is more effective and productive as a means to improve one's state of mind. It also enables a person to achieve his or her goals.

It is crucial to develop skills in mindfulness that will help you in your DBT program. Formal mindfulness helps you focus your attention and improve awareness of situations. For example, you may hold your breath to observe your breathing pattern, as well as what's going on in your body during that time. How fast or slow is your breathing, for one? Similarly, you can go on a long walk into the woods to become aware of your senses. You might also pay attention to the sights and sounds of things that are around you, including the noise that leaves make when crunched under your feet, the rain that is gently dropping on the roof or the chirping of the birds nearby.

Mindfulness can be practiced anytime and anywhere. You do not have to be in a therapy room or with a therapist to practice it. You might be sitting in a meeting and trying to focus, but you are worried about something that has captured your attention. Mindfulness allows you to drown out these distractions and bring your concentration back to the activity that you are doing. It is about noticing the disturbances that you experience and bringing you back to the task at hand. The patients who are doing dialectical behavior therapy should realize that they need to familiarize themselves with such distractions. Nevertheless, they also should know that it is impossible to get rid of all the disturbances; that's why they must learn how to cope with these sources of discomfort and stress.

Tips on Practicing Mindfulness

1. Start thinking about one task at a time. Do not try to do too much at once. You need to be mindful of each task that you are doing to lessen errors and repeating the activity.

2. Practice making observations out loud. If you are looking at things outside, you can be mindful by noticing them and then voicing out your thoughts. It helps you to do a meditative practice of externalizing your ideas.

3. Take note of your distractions. You should notice when your mind starts to drift in a given situation. Be aware of the situation but try not to react to it.

4. Be gentle with yourself. While you begin to practice mindfulness, you need to take care of yourself since it will take some time to develop the technique and have a more peaceful state of being and thinking.

Techniques and Exercises for Using Mindfulness in Your Life

There are several techniques and exercises that you can do to begin your journey towards mindfulness today._27 Positive Psychology Program Your One __ These methods can provide you with ways to relax and de-stress after long days and enable you to feel better about different aspects of your life.

Meditation

This seems like a stereotypical thing to do, but it is useful in getting you to relax and enjoy the present moment. So, take a minute to sit on the floor and close your eyes. Light a candle and play some low-key music that will engage your ears and allow you to de-stress and unplug from your busy life. You can think about different things that will enable you to get your mind off of the stressful circumstances that you are going through.

Guided Meditation Through Audio Recordings

In this kind of meditation, you are guided through various thoughts in your mind. The recording indicates what you should think about as you are meditating, and usually there is music that accompanies the recording. This technique can also be done in a dim room that does not have a lot of noise. You can use some noise-canceling headphones to block out all the background sounds if you wish.

Do the Raisin Exercise

This exercise involves looking at any food, but we can use the example of a raisin. The participant looks at the fruit and describes how it looks, feels, how their skin responds to moving it, as well as the smell and taste of the object. As the person looks at it, he or she can focus on the here and now and not on anything else.

The Body Scan

In this practice, the individual is going to concentrate on each part of his or her body. It is as if he or she is doing a body scan. The participants lie on their backs and sit still during the exercise. In the beginning, people focus on how they are breathing, including the rhythm of inhaling and exhaling. Then, the guru shows the participants how to pay attention to the body until they become aware of all the parts and how they feel. The body scan runs from the head to the neck to the chest and down to the toes and feet. After the learners are done, they can open his eyes and feel very relaxed.

Mindful Observation

In this exercise, you look out a window and observe everything that you see outside. Instead of naming things, such as bird, person or tree, though, you should notice the colors, pictures, and other figures that are out there. You should pay close attention to leaves blowing, the grass swaying in a breeze, and the different shapes of the structures. Try to remain focused on the task at hand, but do not get too distracted by any of them. If you need to take a break, you can get away from the window for a few minutes.

Mindful Listening

Such an exercise is a form of therapy that you can do with other people. You think of one thing that you are stressed or worried about, as well as another thing that you look forward to in the future, and then share it with other participants. The therapist will ask everyone to think about how they are talking about their thoughts. You should observe your thinking patterns and emotions as you speaking and listen to others. Then, you should all reflect on the practice together.

Self-Compassion Break

Self-compassion is a way to cope with life's difficulties by showing kindness to yourself whenever you are feeling stressed or overwhelmed. When burdened by different situations, you can take a break out of your busy schedule, close your eyes, breathe in and out or give yourself a hug or another kind of embrace. Utter some supportive phrases as well, such as, "I know this is hard, but you are doing a good job keeping yourself together. You are awesome!" You have to practice self-affirming language since it will help you get through any hurdle at a given time.

Hold-and-Stretch Exercise

In this exercise, you tense up your muscles and focus on the parts of your body that are causing you to feel stressed or uptight. You gently put pressure on the muscle and concentrate on each part of the body. Then, you release the stress and tension after a few seconds.

Five-Senses Exercise

Through this technique, you look at five different things that you can see, hear, feel, touch, and taste to notice how they appear to you. It is a quick exercise that can bring you to a state of mindfulness in a pinch if you feel down. The five senses can enable you to become aware of your present state within a short period.

Three-Minute Breathing Technique

In this exercise, you spend three minutes looking at how you breathe. During the first minute, you ask yourself, "How am I doing at this moment?" Think about what you are feeling, thinking, and sensing within that period. The second minute should be spent thinking about your breathing pattern. In the last minute, you should focus exclusively on breathing and how it affects the whole body.

Practice Mindful Eating

This method involves concentrating on what you eat. Before putting the food in your mouth, hold and feel it in your hands. Notice the different aspects of it, including its texture, smell, taste, et cetera. Try to eat slowly to be able to do so.

Think About Your Thinking

This exercise requires you to think about your breathing and your body, as well as your thoughts. You should be aware of what ideas are coming to your head and judge them one at a time. If your mind wanders or strays from the task, you can gently guide your mind back to the meditation phase.

Mindfulness Technique for Anger Management

Mindfulness can put your mind at ease when you feel like exploding in anger. With this technique, you can calm your emotions and think clearly (Cullen et al, 2016). First, you should sit in a comfortable position or chair. Make sure that your feet are touching the floor or any flat surface underneath. You should practice deep breathing to be able to fill your lungs with air well, too. Think about a time when you recently got angry, when you were able to cope with a difficult situation. Think about the strong feeling that you experienced at that moment, as well as how your body reacted to this anger.

Notice how different parts of your body are tensed or stressed and observe what is happening. Then, be compassionate to yourself. You may say, "It is natural to be angry. It's okay. This, too, will pass." Then, get rid of your anger by uttering a simple "goodbye." Once you manage to concentrate on your breathing again, you can calm down. This technique should be repeated as many times as you need to. It should also be worked up gradually from milder episodes of anger to more severe and explosive ones.

Stare at a Circle and Reflect on It for a Minute or Two

This exercise allows you to slow down and focus on one thing at a time so that you can concentrate on the circle. Think about your thoughts, reflect on what you are thinking about, and assess its validity and plausibility.

Case Study

Timothy (18) was recently diagnosed with borderline personality disorder. He somewhat had attention-deficit hyperactivity disorder (ADHD) as well. Many times, he would get upset and angry because he was slower than the other students and become distracted easily by his work. Timothy would go into some emotional tantrums at school, too. He would get frustrated at one point and blow up in the middle of class. Often, his teacher would send him out of the classroom because he could not control himself. Later, Timothy's parents took him to a psychiatrist to find out how to treat his condition. The doctor recommended dialectical behavioral therapy to him.

The first thing that the therapist taught Timothy was to practice mindfulness. He learned a variety of exercises. In one exercise, Timothy would stretch his muscles, pause at his desk, close his eyes, and count to ten. He would also focus on his breathing. This way, Timothy could take a time-out whenever he needed to collect his thoughts. As the teenager learned how to do deep breathing and take a pause when he was getting upset, he understood how to successfully manage his symptoms by going to regular therapy sessions with his therapist. Also, Timothy attended group therapy sessions with other young adults who had developed the same disorder. It helped him tremendously in staying on the path to recovery. Amazingly, within one year of being in the program, Timothy got the symptoms of both BPD and ADHD under control.

As you can see, mindfulness is a powerful tool for managing life even for someone with a mental health condition. All of the techniques mentioned earlier are proven to help the person to be in touch with how he or she is feeling about everyday situations and permits him or her to pause and concentrate on one thing at a time. Among the key issues that it examines is how an individual breathes, as well as how his or her body reacts to the stress, anger, or other feelings that he or she is experiencing. By pausing and taking time to reflect, he or she can examine his or her thought processes and practice self-compassion to calm down. These techniques can be exercised by meditating at home or anywhere that a person may be. You can even practice a few of them when you are at work or doing some activity.

Interpersonal Effectiveness

Using Objectiveness Effectiveness (D.E.A.R. M.A.N.)

D – Describe

Describe the event using facts only. Do not use emotions. Let it speak like a police report if you do use emotions. "Patient seemed upset." It works better to sound like a police report if you talk in the third person. However, don't take this habit in the real world. That's just confusing. Don't make a request or "dry beg". Dry begging is saying obnoxiously passive-aggressive things like, "I really need thirty bucks," or, "Wow, that cake looks good. I wish I had some." The best response to dry beggars is: "Yep. You do." Or it could be, "Yep. It is." If they really want it, they'll get around to asking like an adult. It might go like this – say you're from a religious family, and your teenager decides not to go to church. You might reply, "I've noticed you don't like church. Let's discuss the options of staying home."

This is important so that the other party understands clearly what the situation is before you ask anything, entreat, or make an executive decision.

E – Express

Express yourself with "I feel" or other "I" statements. These types of statements help the speaker take accountability and prevent the listener from immediately going into defense mode. Let's go back to the teenager staying home from the church example. Now, you might say something like, "I feel like you should believe what I believe, but I know that you're your own person, separate from me, and I can't force my beliefs on you. I would like you to come to church with us because my worry is you won't be productive at home."

This is important so that the other party understands where you're coming from when you express how you feel about the situation you've just described.

A – Assert

Assert your position by either directly asking for what you need or stating your position clearly. Don't beat around the bush, don't use euphemisms, and don't hesitate to the point of losing the other party's interest. To continue with the example, let's assert our decision for our hypothetical teenager. "I understand that you don't want to come to church with us, and you are old enough to stay home alone. So, if you choose to stay home instead of attending church, you will prepare dinner and set the table and have everything prepared for us to be able to eat when we return, and you will make enough in case we invite people over unexpectedly. If you are unable to complete this chore, and thus, be productive for the whole family while we are at church, you will come back with us, even if you don't believe it."

This is important because ambiguity creates miscommunication in relationships, and that is the biggest source of contention. Be unambiguous. Set the boundaries now. If you're making a request, it must also be unambiguous, maybe even a little lawyerly.

For example, you might say, "Can I please borrow your car from Sunday to Tuesday? I'll return it by 7:00 pm with a full tank of gas and a wash."

The other party might have other caveats. Such as, "Yeah, but it overheats, so don't go over 55 mph, or over 55 miles away. And my tags are expired, so avoid cops. Or renew it for me."

In which case, you might say, "You know what? I can take the bus. Thanks though."

R – Reinforce

Make sure the other party knows why they should grant your request, or acquiesce to your conditions without a fight. "Because I said so" is not a valid reason. Most people reciprocate naturally. In the teenager example, however, it may not be so natural. Let them know that something's in it for them.

You might say something like this, "You get to stay home from church on the condition that you are productive at home. Since you don't like church and I don't like cooking after church, it's a win for both of us."

Or in the example with the car, it might sound like this, "I actually need to drive to a different city for a few days, but I can't rent a car because of (XYZ), so I'll get your car diagnosed for you, and if I can afford to fix the overheating problem, I will. If not, I'll see if anyone else can part with their car for a few days, or find another solution."

In both examples, the other party can clearly see that they have nothing to lose by accepting your request, and everything to gain.

This is important because relationships are built on reciprocity. When one party feels slighted occasionally, it's not a big deal. But if one party feels slighted more often than not, they will most likely end the relationship.

M – Mindful (stay)

Stay focused on the conversation. If you're answering a text, they have no reason to listen to you. If they're answering a text, that's out of your control, but you can keep your mind on the conversation instead of what they're doing. If they become defensive, notice what you may have said wrong, apologize if necessary, even if it's just to get them back on track.

This is important because it's too easy to go off track and lose focus, especially in an uncomfortable situation, where the other party might be looking to pick a fight. If you go off on tangents, whether they be to sing and dance because some said a song lyric, or to fight, or because one of you saw a squirrel, you have less of a chance of getting what you want. Especially if you're the one singing and dancing or chasing squirrels.

Let's go back to the hypothetical teenager.

Your teen may interrupt you to tell you that they've been cutting church every week with their friends from Sunday School anyway, so there's no point in going. You may have to repeat yourself a few times, especially if you're letting them stay home as long as they're productive, as they may not believe their ears.

Again, repeat yourself as often as necessary, and if you have a real kid, you've done that a few times already this morning. And bring the conversation back to the topic. Detour... focus. If we're using the example of asking an adult friend for something, you don't have the clout you do as a parent. You still may have to repeat yourself, but the interruptions might just be singing and dancing.

You might say something like, "I understand you don't like church, and you cut Sunday School anyway. But you get some out of it every week if you continue going, and I would like that." Or, "If you're going to stay home, you'll need to cook for us, and I'll take the added precaution of changing the Wi-Fi password every Saturday to make sure you'll be productive. If you can prove that you are, that you don't have friends over, that you cook and clean as you go, I'll stop doing that."

A – Appear Confident

Appear confident no matter how you actually feel. If you have this look all the time, little old ladies will ask you for a napkin at a restaurant when you're on a date, and it might not even occur to you to tell them you don't work there, so you walk into the kitchen and get the napkins.

Your nonverbal cues indicate confidence more than your verbal cues. Sit with your back straight and your head held high. Make eye contact. Orient your feet towards them. Where your feet are oriented is where your mind subconsciously goes. Appear confident and stand your ground.

This is important because confidence signifies that your request isn't too difficult to grant and that you're harder to turn down. There's no need to be overbearing like Ferris Bueller. If they do refuse you, in an adult to adult conversation, you might just ask if they're sure, then thank them for their time and let it go.

If our teenager refuses you, this might be a good time to tell them what the other option is. "Okay, you don't have to learn to cook. And if you can read, you can cook, by the way. You can keep coming to church with us, and thank you for letting me know about cutting Sunday School. I'll be sure to tell your friends' parents you all do that because they'll want to know too. I'll let them know you told me. Thank you for caring about the salvation of your friends, who also should go back to church." This will most likely ensure you an excellent meal every Sunday.

N – Negotiate

Negotiate. Remember, "give to get", as selfish as that sounds. Everyone wonders, *What's in it for me*? You aren't demanding something. You're asking for something or setting down a rule. Even in setting down rules, you aren't demanding. If you think you can demand something of someone, even a child, expect defensiveness and confrontation. You give options.

You may need to alter your request to make it more pleasing. In the borrowing the car example, you offered to get the car diagnosed (AutoZone does it for free) and fix it if you could – and if you couldn't, you'd find another solution to our problem.

This is important because building relationships may or may not be the most important reason we spend a few decades on this planet, but it certainly takes up most of our time. Whether we spend that time in actual relationships with other human beings or wondering why we drive away other human beings, we spend an inordinate amount of time either with other people or thinking about them, whether we know them personally or not.

So if we spend our energy browbeating others and expecting them to kowtow to us, that only works if you have money, and even then not everyone likes you, even if you somehow win elections. Mere mortals, without insane amounts of money, can't behave like that. We have to negotiate and play nice.

Going back to the example of the teenager, this is pretty much already a negotiation. They still refuse to go to church or cook, and they tell you they don't care if you call their friends' parents. They really do. This is when you pull out your phone and look up the numbers of the kids' parents, who you probably know, at least by name already. Google white pages are great. Some rules are not to be negotiated. If, however, you start the conversation with trying to force them to go to church, this idea is a perfect negotiation, and now it seems (to them) that you've given in some.

But for example's sake, you do try to negotiate. You might say, "Okay, if you don't come to church with us on Sunday, you still need to be productive at home. Would you rather have a list of chores to do? What is your suggestion to being productive, other than homework, because I don't want you deliberately putting it off until Sunday?"

This approach helps your child feel like they have a say – like their voice is heard and not invalidated. If you start off demanding they go, then negotiate to this, you can offer it as a suggestion, and ask which of your suggestions they like best.

You can both leave the conversation feeling like you've accomplished something, like you've got a win, like you're helping the other person out, with no ill will.

D.E.A.R. M.A.N.

Using the DEAR MAN skills will improve any relationship. Start now. If you're yelling at each other all the time, at least you're still talking; there's still hope. It's when you've stopped talking entirely that it's harder to reconcile. These skills can reduce arguments and improve understanding. Let's put the whole conversation together now.

Describe: "I've noticed you don't like church. Let's discuss the options of staying home."

Express: "I feel like you should believe what I believe, but I know that you're your own person, separate from me, and I can't force my beliefs on you. I would like you to come to church with us because my worry is that you won't be productive at home."

Assert: "I understand that you don't want to come to church with us, and you are old enough to stay home alone. So, if you choose to stay home instead of attending church, you will prepare dinner and set the table and have everything prepared for us to be able to eat when we return, and you will make enough in case we invite people over unexpectedly. If you are unable to complete this chore, and thus, be productive for the whole family while we are at church, you will come back with us, even if you don't believe it."

Reinforce: "You get to stay home from church on the condition that you are productive at home. Since you don't like church and I don't like cooking after church, it's a win for both of us."

Mindful (stay): "I understand you don't like church, and you cut Sunday School anyway. But you get some out of it every week if you continue going, and I would like that." Or, "If you're going to stay home, you'll need to cook for us, and I'll take the added precaution of changing the Wi-Fi password every Saturday to make sure you'll be productive. If you can prove that you are, that you don't have friends over, that you cook and clean as you go, I'll stop doing that."

Appear Confident: "Okay, you don't have to learn to cook. And if you can read, you can cook, by the way. You can keep coming to church with us, and thank you for letting me know about cutting Sunday School. I'll be sure to tell your friends' parents you all do that because they'll want to know too. I'll let them know you told me. Thank you for caring about the salvation of your friends, who also should go back to church."

Negotiate: "Okay, if you don't come to church with us on Sunday, you still need to be productive at home. Would you rather have a list of chores to do? What is your suggestion to being productive, other than homework, because I don't want you deliberately putting it off until Sunday?"

Using Relationship Effectiveness (G.I.V.E.)

GIVE is the DBT acronym used for obtaining and maintaining relationships. This is difficult for many people, but more so with people living with mental illnesses.

G – (be) Gentle

Be kind when addressing people. It seems we're much more capable of being kind to complete strangers and people we work with (but don't really like) than people we actually love. If you have a difficult topic to address with a loved one, be conscious of your tone of voice, facial expressions, body posture, and other nonverbal cues we all send every day.

Be able to take NO for an answer from the DEAR MAN approach. Again, this is a conversation between adults and not your hypothetical child. They have every right to say no without being guilted, judged, manipulated, or begged.

Don't leave the discussion or threaten or manipulate them. Don't threaten to hurt them or someone else or yourself. If you're truly feeling the need to self-harm when talking with them, apologize for trying to start this discussion, leave the situation, and seek help.

Don't name call, condescend, bring up past injuries or "debts" you think they owe you. They do not owe you the right to say yes, no matter what you've done in the past. If this is a one-sided relationship, you may want to reevaluate how much time and energy you invest in it. Don't guilt trip or force the issue.

Thank them for their time. Leave the situation or change the subject at an appropriate time. Remember this: no one owes you anything, and you don't owe anyone anything. Be kind. Use your inside voice, and go right on back to being friends after they say no. You're an adult. You can solve this problem.

I – (act/be) Interested

The manual says to act interested in what the other person has to say; however, you should instead *be* interested. Anyone can act interested. You're probably acting interested now, reading this. But in discussions with loved ones, especially difficult discussions, the kind that a DEAR MAN approach necessitates, you need to be interested. They're going to say something that may be important in a later conversation.

Be interested. Be present. Be mindful. Show them the same interest you would like shown. Don't invalidate their words by your words or actions (or inactions or silence). Let them feel heard. Really listen to what they're saying, instead of waiting to speak. Worse yet, don't interrupt them to speak.

Ask open-ended questions that show you're actually listening. Pay attention to your body language and make sure it's inviting and also showing interest, like nodding. Maintain eye contact. When you do speak, make sure you're not speaking only about yourself. Make space for other people in the conversation. Be comfortable with silence. Really, it's okay.

Be cognizant of the other person's needs. If they wish to discuss something else, make sure this topic is done, and both parties feel satisfied. If you're not there yet, let them know you're happy to discuss it at another time.

V – Validate

Allow the other person to feel what they feel. Don't tell them how they should or shouldn't feel. Don't tell them they're just talking like a debater when they make good points – as if that fact invalidates their argument. Don't dismiss what they have to say because they're younger than you or because they haven't had your experiences.

Let them know you understand where they're coming from, even if you don't agree with their point. "I totally understand why you think that now. Thank you for explaining it like that. I still don't agree, but I'll stop trying to change your mind on the matter."

Try to determine what part of your request the other person is uncomfortable with. "I know you're busy, but I'm not in a rush." Or, "I didn't know your car was such a piece of crap, and that's why you work from home. I'll figure out a way to rent a car."

Even without conflict, this is an important interpersonal skill to learn. Everyone likes to feel validated and heard. Everyone likes to know they matter in a relationship. Validating other people is one of the most necessary facets in relationships of trust.

E – (use an) Easy Manner

Smiling goes a long way in conversations with or without conflict. Not the forced, "I actually want to kill you" smile, but a real one. Unless you've worked in customer service for upwards of a decade, that smile always looks fake. If you have, you're a natural. You can give fake smiles to the best human lie detectors around.

A little humor goes a long way too. If you tell a joke and it falls flat, blame it on your dad. "That was one of my dad's worst dad jokes. Actually, it might have been his best. He's an accountant." If you say something serious and they all look at you weird, then you were kidding. "I'm kidding! When you look at me like that, I'm kidding!"

Offer a sincere compliment. If you can't think of one, tell them you're jealous of their clean criminal background. If you don't have a follow-up story for why you'd say that, you were kidding. When they look at you like that, you were kidding. Lighten up the situation with a little self-deprecating humor. Really, that stuff is the best. If nothing else, just smile and nod and give them a little more attention. Make them feel special.

Self-Respect Effectiveness (F.A.S.T.)

F.A.S.T. is about maintaining self-respect during conflicts. You may have had the role of doormat for way too long. You may have the role of the youngest, the "baby", and thus, always treated as such, no matter how old you are. Set healthy boundaries, even late in the game. Maintain your self-respect. This is possible even with the self-deprecating humor if you've set healthy boundaries. You'll want to use these skills in sequential order, then altogether.

F – Fair

Be fair to yourself and others. Those with borderline personality disorder usually also have anxiety and depression (mental illnesses are like potato chips – you can't have just one). One of anxiety's favorite past times is to make every decision, action, reaction, everything that happens, completely catastrophic, apocalyptic, no coming back. It's great fun. Here's a little trick. Name your anxiety. Use a fictional character you hate. If you use a real person you hate, it'll increase your anxiety. When you get these crazy end of the world scenarios building up, say, "Umbridge, go away." Or whatever you named it. That helps to separate yourself from the anxiety and helps you recognize *you're* not freaking out - your anxiety is.

Be fair in both thoughts and actions. Instead of thinking the other person is the worst person on the planet, or you're the dumbest person ever born, redirect your thoughts to something fairer, then focus on the subject again. Find the elements of truth in their argument. Ask clarifying questions if you can't. Get to a point where their argument makes sense to you. You don't have to agree with it, but now you understand it, and you may even agree with elements of it.

A – (no) Apologies

This isn't to say don't apologize when you're in the wrong. When you're in the wrong, absolutely apologize. This isn't to say don't apologize when you're not in the wrong. A common example of this is a typical woman. She'll say "Sorry" as she passes you instead of "Excuse me," or "On your left."

When you're resolving a conflict with someone, you might be tempted to apologize just to get it over with and accept blame when it doesn't belong to you. Maintain your dignity. Don't do this. Don't even offer a fake "I'm sorry you feel that way" apology. Offer a "Why do you feel that way?" or a "What responsibilities do you think you have in this scenario?" open-ended question without appointing blame or getting defensive.

S – Stick to Your Values

To stick to your values, you first have to know *what* your values are. Determine your values before opening this dialogue. Do some introspective self-examination and figure out what's important to you, what you won't bend on, and what you can bend on without losing any respect for yourself. The only person who HAS to live in your head is you. Make sure that is a person you can respect and love. Be honest about what you say your values are, and what they actually are. If you say you value fiscal responsibility but never budget, you need to either start budgeting or stop saying you value fiscal responsibility. If you say you value family but cheat on your wife, reevaluate yourself. It might help to write down your values then write down the actions you do that would corroborate those values. This might help you to determine your values. After you know what you stand for, keep standing.

T – (be) Truthful

Be truthful with yourself and others. Do you downplay your accomplishments and exaggerate your faults? So you downplay or avoid your faults entirely and exaggerate your accomplishments? Do you take credit or blame when it belongs to someone else? In the adult world, there is no need for this. You will be viewed as more honest among others if you can be honest with yourself. Are your words true? Will your version of a story closely match other people's? Are you as objective as possible about yourself and others?

Using these steps will allow you to maintain your dignity. You may not be pleased with the outcome of the discussion, but you can walk away with your head held high.

T.H.I.N.K.

These are the basics of Dialectical Behavior Therapy. Using these steps will help in interpersonal relationships with or without conflict. There is a newer acronym, not widely known, but it is useful in interactions with conflict. It's called THINK, and its purpose is to help you reduce negative emotions towards others.

T – Think

Think about the situation from all angles. Think about it from the other person's point of view. Try to step into their shoes. Do they view you as unreasonable as perhaps you view them? Listen to the words coming out of your own mouth with an objective mind.

H – Have empathy

This is where you really try to walk around in their shoes, as Atticus Finch would say. What else is going on in their lives that might make them behave like this? Their behavior towards you might not be entirely about you. What does it feel like to be them? Let yourself feel their emotions for a moment.

I – Interpretations

Interpret and misinterpret their behavior. Try to figure out why they might be behaving this way. Start with outlandish ideas to firstly open your mind, and secondly, realize how outlandish it sounds so you can come close to what might be a reality. For example, you might start out with this assumption and lead yourself to less outlandish ones:

She's a test subject to see how mean she can be and is getting paid for this behavior. She's a test object to see how mean people can be to her, but it's a double-blind study, so she doesn't know why she's getting paid or why a particular person is horrible to her, and she's reacting to that. She's not involved with any tests about meanness but just enjoys doing it for free. Her kid got called to the principal's office again, and she's sick of missing work for him and doesn't know what she's doing wrong. She's tired and having a bad day, bad week. I should just tell her she's doing a great job and not let it get to me.

The point of that exercise is to bring your anger down enough to be able to do the next two steps.

N – Notice

Notice the other person. Notice if he or she looks tired instead of angry. Notice if he or she has a lot on their plate. Notice if you are inadvertently adding more. Notice if they smile at you, even if you're not on good terms. Hopefully, you smile back, but you don't have to do anything just yet. Just notice.

K – Kindness

This doesn't mean forgive and forget. This doesn't mean that if they apologize, you can go back to treating them poorly – nor can they do that to you. It just means be kind. You may need your space for a bit. Instead of saying, "Leave me alone! I don't want to talk to you!" try saying, "What you said really hurt, and I need some time to process it – so I need some space for a little bit, please." This will give you a much better response, and when you're ready to talk to them again, they'll be more willing to express in a civil manner why they said what they said or did what they did – and maybe they'll apologize, and maybe they won't. In the end, what's important is they will have nicer feelings towards you.

This skill set could be determined as an interpersonal distress skill set. Like all the acronyms we've discussed, you won't become a master of any of these overnight. They all require daily practice and self-reflection.

There's one more thing to add to this extensive list of improving interpersonal relationships. It's not an acronym, and it's not found in any text read during a master's degree in interpersonal communications.

It's **T.H.A.N.K.** Just thank the other person for something every day. It might be small. It might be something they should do because they live there, too. If, for example, your husband takes out the trash every day, start thanking him. If your wife is a stay-at-home mom, start thanking her for the million thankless things she does every day. A simple act of gratitude will get you far, even just something like, "Hey, I don't know what your job entails, but I know it's hard – so thanks for letting me work. Thanks for keeping the kids alive and fed every day."

Distress Tolerance

There are people who can overcome a crisis remarkably and there are those who are so overwhelmed by their emotions that they become debilitated.

Those unable to withstand the crisis resort to self-defeating escapes from the pain like substance abuse and self-harming behavior. This is often the case with people that have borderline personality disorder.

The distress tolerance module in DBT teaches people to control radical emotional suffering induced by crisis. By building their resiliency and showing them various avenues to deal with difficult situations; it helps them adapt an effective coping strategy when dealing with distressing situations. Distress tolerance skills training equips an individual with the mechanics of tolerating emotional anguish and have him/her avoid self-destructive paths.

Distress tolerance skills give the participant short-term strategies to deal with harrowing situations. There are two kinds of methods:

Acceptance: This is a non-judgmental acceptance of the situation and oneself. This strategy accepts emotional suffering as an inseparable element of life and believes that failing to accept this actuality brings about further suffering. Breathing and awareness exercises are methods used to teach acceptance.

Crisis: These strategies are aimed at finding new ways to manage and survive without turning to self-defeating avenues. In DBT, participants are taught four crisis survival skills: contemplating the pros and cons, improving the moment, self-soothing and distraction.

Distraction

The techniques of distraction can easily be memorized through the acronym **ACCEPTS**.

- *Activities*: You can easily distract yourself by indulging in pleasing activities like gardening, playing sports and walking. Jotting down a few of your favorite things to do can serve you during crises.

C- *Contributing:* This is about drawing your mind away from yourself. You can enroll yourself in voluntary work or do something nice for a family or friend. It will have you think of the person or the charity work and not yourself.

C- *Comparison:* This is a strategy in which you start comparing yourself with others who are striving to cope the same way. You can also compare the level of progress you've made since you first started. You could do this by watching movies or reading books about tragedies. This will give you a clear viewpoint of the distressing situation and help you see that there is worse out there.

E- *Emotion:* Here you are encouraged to work contrarily to your emotions. For instance, if you feel depressed and you just want to lie in bed all day, you are urged to go out for a walk or engage in a physically elevating activity. This is "opposite-to emotion action"; it has been discussed earlier.

P- *Pushing away:* This is about postponing the thought of a distressing situation. Imagery and practical exercises are implemented. For example, you can put moments of turmoil in a box and leave them there until you feel emotionally fitted to address them.

T- *Thoughts:* You are taught to distract yourself from negative or distressing thoughts. Reading, watching movies, writing a diary, counting from one to ten are some of the methods employed for this strategy.

S- *Sensations:* Through the use of techniques like taking a cold shower, holding ice in the hands, listening to loud music and flicking a rubber wrist band, you are taught to awaken the senses which will briefly distract you from emotional chaos, opening the door for the other distraction techniques.

Self-Soothing

It's quite impossible for those who love and care about you to always be around when you need them in your moments of distress or agitation. You need to adopt a survival skill that can help you overcome such challenging moments in life.

In DBT participants are taught self-soothing skills, which is the skill to nurture, comfort, and be kind to oneself. It is a fairly simple skill to adopt but for those who have suffered from Borderline Personality Disorder, it entails certain challenges.

People suffering from BPD tend to seek out validation and comfort from other people and so this independent way of curing their own ills often has a flustering effect on them.

Self-soothing skills in DBT are taught by using the five senses: **Touch, Hearing, Taste, Smell and Vision.**

Vision

There are so many beautiful, captivating things around us that we can use as a distraction during tough moments. Pay further attention to those little things; be it a wall, a colorful tile, a tree, birds, moving vehicles, flowers; and study their features as much as you possibly can. This visual distraction will give you the break needed to recollect your thoughts.

Hearing

By paying attention to the sounds around us, we can easily calm our nerves down and evade the overwhelming emotions that distressful moments ensue.

So if classical music, hip-pop, rock, or rap has that tendency of calming you down then download it on your phone or iPod and listen to them at such stressful moments. And if the situation doesn't permit you to listen to music, then listen to the sounds outside; pick one sound that captivates your interest and focus on it. Or you can just attentively listen to your own breathing.

Smell

Nothing can trigger a memory as effectively as that of odor. Your mother's perfume or your father's cologne, for instance, is unforgettable. And by focusing on those memories that the odors bring about, you can sooth yourself effectively. So go out shopping for a perfume, lotion, or deodorant that can take you back to the Eden of your childhood.

Or you can mindfully smell whatever kind of odor is around you. If, for instance, you are taking a shower or bath, take a moment to smell the soap or shampoo. Or when you are dinning, embrace and savor the essence of your meal.

Taste

Things that have a pleasant taste have a soothing effect. Have a small bit of a chocolate bar or mint and try to stimulate your taste buds. Savor it in your mouth first then focus on how your body reacts to the taste.

This is an effective method necessary to distract yourself when you're in some distressful situation. It can, however, be a challenge to those struggling with eating disorders or those who have an emotional attachment to food. So it's highly suggested you use a small chocolate bar, a peppermint, or a gum for this practice.

Touch

Some people find their utopia in messages and others in bubble baths, a clean sheet, or silk. So what is yours? Explore and identify the kind of touch that has a calming effect on you then treat yourself with it.

Improving the Moment

When overwhelmed by strong emotions, it's quite impossible to keep anything under control, let alone a crisis. But by staying mindful and improving the moment you would be able to tolerate the inundation of negative emotions during distressful moments.

This skill can effectively be put to practice and remembered by the acronym **IMPROVE**.

I – *Imagery:* Positive imagination is a very powerful tool that can help you keep strong emotions under control.

Picture yourself in a distressful situation; alone and deserted on an island or losing a job; then imagine the positive aspect of the scenario, which is you overcoming the crisis. This technique will take you to a safe and relaxing state of mind.

M – *Meaning:* To every downside there is an upside. Painful and stressful experiences always have a thing or two to teach us; we just have to look deeper into them. And by giving your situation a certain amount of value and meaning, you can improve the moment and overcome your throb-inducing emotions.

P – *Prayer:* You don't have to be a devout disciple of some religion to use this technique, all you have to do is believe in a higher being, be it God, the Buddha, or your own Wise Mind, to whom your prayers are directed.

There are two kinds of prayers in DBT; one is where you ask 'Why me? 'In this prayer you ask the numinous entity if there is a reason, a purpose behind the pain you're experiencing in life.

And the second is where you ask for the bestowal of strength and relief.

R – *Relaxation:* To improve the moment relaxation techniques can be employed. In this practice, methods such as deep breathing, listening to music, taking a bath or progressive muscle relaxation can be employed for relaxation.

- – *One thing in a moment:* This is a technique that has much semblance to that of the teaching in Core-Mindfulness skills. By focusing on just one thing at a time, you'll be able to give yourself enough leeway to calm down and deal with crisis accordingly.

V – *Vacation:* Vacationing DBT doesn't quite necessarily mean 'going off to a holiday'. It is a metaphorical expression that basically means 'Give yourself a Break'. Grant yourself an entire day off and engage in an activity that entertains you best; read a book, go spend some quality time with your friends or relatives, treat yourself to a spa or your favorite food or drink. These activities will allow for a break from the crisis.

E – *Encourage:* When you are amidst some stressful situation, have a positive conversation with yourself. Acknowledge your efforts and encourage yourself by repeatedly saying things such as 'I can do this', 'I am doing the best I can', 'this will not go on forever', 'I will be okay'.

Remember that for any of these skills to work effectively, you need to practice.

Pros and Cons

This is a part of the distress tolerance module where the individual is asked to jot down a list of the disadvantages and advantages of tolerating and the same list for not tolerating distressful situations. He/she is asked to evaluate the consequences of potential choices. This exercise aims at having the individual focus on the long-term goals and benefits of overcoming a crisis.

Let's assume that you just got informed that your dance class was cancelled. Tolerating the distressful situation would be you accepting that you are upset, but able to understand that you'll attend the class next time. Failing to control the anguish would be you lashing out at the tutor and assuming a vituperative stance.

The consequence of your first action would be a brief moment of anguish; however, you would be able to go to your next dance class. The consequence of your latter action would be a feeling of shame after such an unreasonable reaction and a potential expulsion from your class.

Eventually, you learn that tolerating distress instead of acting on impulsion is a better option in the long run.

- **Radical acceptance**

Radical acceptance is about accepting the situation and ceasing to fight, accepting the situation as it is rather than how you want it to be.

Radical acceptance believes that moving on is only possible when one acknowledges the fact that the past is unchangeable.

Your mind will work effortlessly to prohibit you from seeing a situation you are in as it is rather than what you want it to be. Your mind will fall back into old patterns giving you other interpretations, thoughts and ideas of the situation that are foreign to the reality. To stop this you will have to fight off these disturbances and maintain focus on the moment at hand. This skill is called "Turning the mind."

Radical acceptance is not about deeming the situation as good or bad, allowing the prolongation of the situation or forsaking your options. It is about believing that everything is "as it is" and "as it should be." It is about accepting what is.

The individual is taught to focus on the present through the skills of mindfulness and distress tolerance.

How to control your emotions - Emotional Regulation

When normal people are born, they usually have almost no emotional intensity, but then their emotion regulation rises up a bit when they are bullied at school. When they experience a breakup at secondary school, their emotion regulation rises up even further. Later in life, they get a sack letter from their workplace, and then their emotional intensity goes up a little bit.

When they get a divorce, it goes up a bit further. Their emotional intensity continues to rise until the moment that it reaches a breaking point. The breaking point is the point where people attempt to kill themselves, participate in self-harm, or take harmful drugs.

However, for people suffering BDP, they're born with a lot of emotional intensity. They go through life the same way that a normal person would go through, but at each stage of their life, their emotional intensity increases and refuses to come back down like it would for a normal person. That's why they reach the breaking point earlier in your life and begin to have all strong impulsive emotions to commit suicide.

So that is why you need to learn how to regulate and manage your emotions instead of allowing your emotions to manage you. What you'll find is that you'll have intense emotions that change frequently. Your emotions are what drive your behaviors. It's your emotions that make you do the things that you do.

Sometimes, your behavior comes as a result of trying to get some validation from your emotions or trying to block your emotions or run away from them because they're so painful. So in this section of the DBT technique, you'll learn the skills to successfully manage your emotions and how to become less vulnerable to negative emotions. You'll also learn what emotions are and why you have them. Not only that, you'll learn the techniques for increasing positive experiences and emotions because your life right now is filled up with too many negative emotions. You'll learn how to decrease your emotional suffering. You're not going to invalidate your emotions because your emotions are very real and you have to believe and accept them. You are not going to learn to avoid them or dismiss them. Emotions are okay, even negative ones. The problem is that they cause a lot of disturbance and negativity in our lives. That's why you must learn to manage them. This emotion regulation is not here to dismiss or invalidate your feeling even though other people are currently invalidating your feelings and telling you to get over it. They are telling you that just to cheer you up and make you feel like you don't have it that bad, but deep within you, you know that your feelings are very real to you, and they're there for a reason, even the bad negative feelings. So here you'll learn to look at those feelings and accept them the way they are. Your behaviors are driven by your emotions. You're either behaving in a certain way to get your emotions validated so that people know how you're feeling or you're overdrinking to hide your emotions.

Using Opposite Actions

One of the emotional regulation skills that are very effective for DBT is 'the opposite action'. Opposite action is basically what it says. It means doing the opposite action of exactly what you're feeling or acting opposite to the way that you want to act. You can use the opposite action when you're experiencing emotions that don't fit the facts. That's not to say that your emotions aren't valid enough. Of course, they're valid, but they might not be at the intensity level that makes sense for that situation. You can also use the opposite action when you're having urges and want to act on those urges. Acting on those urges is not going to be effective. If you're feeling really angry, you might want to punch someone in the face or start a fight. The opposite of that will be not to fight but to walk away from the situation. Now, doing the opposite action doesn't mean that you're suppressing or invalidating your emotions. You're still going to feel the emotions, but now, you're going to be doing the opposite action of what you're supposed to do. Then overtime, doing the opposite action of your emotions will help to change those emotions. It'll give you a better outcome because if you punch someone in the face, the outcome will never be as good as if you had just walked away and calmed down. The opposite action walks for different kinds of emotions. If you have terrible anxiety, that's making you so scared of leaving the house, then the opposite action of that will be to face your fears and just go out. It may not be easy to do, but it'll keep getting easier once you've done and achieved it the first time. It'll keep getting easier and easier to the point where you can go out without that fear of anxiety.

If you're depressed, the normal thing for you will be to refuse to get out of bed, avoid seeing anyone, or refuse to do anything. Because, when you're depressed, you'll feel so tired and drained to the point where you just want to lie down and be left alone. So the opposite action will be to go for a run, go to the gym, or have a nice energetic workout. Of course, it won't be easy to do the first time you try to go for a run, because, when you're in a depressed state, the last thing that you'll want to do will be to go for a run, go to the gym or do something energetic. However, if you manage to do it, you'll find out that you don't feel depressed anymore. You'll feel good because the exercise will boost the dopamine, and serotonin in the brain (The feel-good chemicals). Even if you can't go for a run, just get up and go for a walk. Do anything to get out.

If you're feeling sad, the normal thing to do will be to cry your eyes out, but then the opposite of crying will be to do something that makes you laugh. Although there are certain times, where it's good to let yourself cry out loud so that you can feel better, when you find yourself crying constantly, you need to do the opposite action of crying, which is to laugh. By doing the opposite of the negative emotions, you're still acting on the emotions, but you're doing so in a different way, and the outcome won't be negative, It'll be positive.

The 3 Steps for Taking Opposite Actions

The first step of the opposite action is to **Identify the emotions**. For instance, let's say that you're experiencing anxiety before writing a test. The first step is to check the facts about the anxiety that you're feeling. You can tell yourself that you've studied already even though you know that you could have studied more. You tell yourself that you've been doing okay in other tests, and It's not a midterm exam, not a unit test. After checking the facts, you'll notice that you're not supposed to be feeling as anxious as to how you're feeling.

The second step is to **identify your urges**. In the case of the anxiety, your urges might be to skip school, fake a sickness, call ahead of time to inform your professor that you're not taking the test. However, if you closely look at all these urges, you'll see that none of them are going to be effective in dealing with the situation. Even though they take you out of the test, they just push the test to a much longer date, which will likely increase your anxiety even more. So you have to understand that that action urges are not going to be effective.

The third step to using the opposite action is to **Ask your wise mind** -which we talked about before as the combination of your emotions and logic. In the case of the test, the opposite action will be to go and take the test. So you use your wise mind and ask yourself, "what is the most effective thing for me to do at this moment." After asking yourself this question, you'll notice that the opposite thing for you to do at that moment might be to go to school, talk to your professor about it, talk to your guidance counselor about it, and do anything that'll be effective in taking down your anxiety in the long term. So once you've identified the opposite actions, and you've figured out which one you're going to take, proceed and do that damn thing. Go to school and take that test. Do the opposite action of the thing that you want to do now. To make your brain to know if taking the test if the right action to do now, Just go take the damn test. Then after taking the test, you can now tell your teacher about your anxiety on the test, and that you'll like to know what you can do about the next test, so as not to feel really anxious again. You could ask your professor to cover things for you in those chapters that you're not really comfortable about. So go the extra mile about your anxiety, so that next time your anxiety levels will come down and everything will start to click in your head. If you try the opposite action the first time, and it didn't work, you can simply go back to your wise mind to see what you can do next and then just repeat acting opposite until it clicks, and you find the opposite action that makes sense to you.

You also want to use the opposite Action when the intensity of the emotion you're exhibiting does not fit the facts of the situation or when you're exhibiting an emotion to a situation for a longer period than you're supposed to.

When using the opposite action, you need to identify the emotions that you intend changing. And then check to see if the emotions that you're exhibiting fit the facts of the situation or not. If the emotion does not fit the facts of the emotion that you're trying to change, then you should try to identify the action urges. For instance, If getting angry does not fit the facts, then you want to identify the action urges which might be to scream, yell, throw things, or raise your voice. So when you feel that emotion, identify the behaviors that you want to do. Next, you want to act the opposite of each emotion. So you want to repeat the opposite action event continually until the tendency for you to follow through with your action urges goes down, and your emotional sensitivity goes down. So continue to practice the opposite action as long it takes to lower your emotional response. Over the long hall, just keep practicing it repeatedly, until eventually condition yourself to act more effectively. So let's go over the opposite action that you'll use for the different emotions. Let's go over some emotions and the opposite action that you can use for that situation, let's start with fear.

Fear: When fear does not fit the facts of the situation, then the opposite action is to do what you're scared of doing over and over again. If you're afraid of going to your exposure group, then the best way to act opposite to that will be to go to your group over and over again. Another way to deal with the fear emotion is to approach people, activities, tasks, and events that you're scared of and also practice to gain a sense of control and confidence over your fears. The most important point to note while doing the opposite action of fear is to keep your sight and hearing really focused on the event. Be aware that you're still physically and mentally safe, even though you're afraid of exhibiting fear right now. So while doing the fear-based opposite action, try to change your posture and let your tone of voice remain confidence. Do pace breathing and try to breathe more slowly and deeply into your diaphragm.

Anger: when your anger does not fit the fact of the actions, the opposite action is to avoid the person that you're angry with rather than throwing an attack on that person. Now gently avoiding is different from fully avoiding, because if you're feeling like yelling or shouting at the person, then you want to avoid the person until you calm down. Do some little paced breathing, and when you become a little bit calmer, you can come back to the person. You can also decide to take some time out to breathe, deeply and slowly. People with BPD tend to react strongly whenever they're angry. So you need to take a moment to breathe in slowly until you get calm, then you can come back and start trying to communicate again.

Another thing you can do when you're angry is to do something really nice to the person rather than pouring insults on that person. For instance, when you call a customer hotline number, you might have the urge to get utterly disgruntled. So instead of been mean and rude to the customer hotline, think of the fact that if you want someone to help you, you have to be his or her friend rather than their enemy. So you should try to be a little bit nice to them instead of reacting to your urge to get upset and yell. Instead of yelling, you can say something like, "I know this isn't your fault that my acct is this way, but I don't know if you can help me.. ". When doing the opposite action for your anger, try to understand and empathize with the other person. Also, try to see the situation from the other person's point of view. For instance, in the case of the calling customer care that we talked about earlier, the person on the phone may be really putting off and not paying attention to what you're saying because they lost someone and still had to come to work. So trying to see this from this other person's perspective can help you to take the opposite action. Always imagine possible reasons for why they're behaving the way they're behaving. In the customer care example, for instance, it may also be the person on the first day at work, and he might not be used to talking to customers on the phone; that's why he's really nervous. They might be many reasons for someone to act the way they're acting. Another way to act opposite to anger is to use the anger energy to do a workout. You can also pour cold water on your body to help change your body temperature both physically and figuratively in your mind, where you're really hot and angry.

Disgust: when disgust does not fit the fact, the opposite action will be to eat, drink, or move closer to the thing that you're feeling disgusted about. This is particularly helpful if you're dealing with people from other cultures and races. Approach them, try to be near them and walk in their own shoes so that you can get to know them as just humans. Also, try to distract yourself from any disgusting image or thought in your mind. If something is really grossing you out, then think about something else.

Envy: when envy does not fit the fact, the opposite action will be to stop yourself from destroying what the other person has accomplished. For instance, if you're in a competition that you got first runner-up, instead of trying to ruin the experience for the other person by talking behind their back, do the opposite action of the envious urges. So if your urge was to bring the person down, or say negative things about them, then you should congratulate that person for doing a really great job. Also, be grateful for the things that you've accomplished. Be grateful for being the first runner-up out of the many people that participated in the competition. Try to look at the things that you do have and be thankful for them and avoid discounting some blessings. Don't discount some blessings because you have the urge to be envious. Avoid exaggerating your deprivations. Also, don't be too harsh on yourself or make comments like, "I am such a loser, I never win anything." Stop exaggerating other people's net worth or value by saying things like, "They have everything, while I don't" the fact that someone is Rich doesn't necessarily mean that another aspect of their life is also balanced, so check the facts. Instead of exaggerating the other person's Blessings or discounting it, check the facts to see why the other person has what you don't have. Maybe the person worked a little harder than you to become rich, or maybe the person entered the competition a little bit earlier and had more time to prepare.

Jealousy: when jealously does not fit the fact, the opposite action is to let go of trying to control other people's actions. If you're the type of person that likes to check your loved one's messages, then that is something that you have to let go of. Share the things and people in your life with other people. If you're not comfortable with your loved one spending their time with other people, then let them spend their time with other people. Don't spy, or snoop on other people. Don't read their emails, or use their passwords to look at their stuff. Also, stop asking to suppress probing questions like, " who were you with? Why were you there? How comes you were talking to him or her?" If you have the urge to be jealous rather than going crazy over the situation, try to calm down and listen to all the details of what's going on. Also, keep your eyes open, be mindful, be aware, and pay attention to the facts of the situation rather than listening to your emotions. Consume all the information about the thing that you're trying to be jealous of. Don't just gravitate towards jealously; try to gravitate towards the things that you don't think about when you get jealous. Look at the part and piece of the situation that doesn't come naturally to you. For instance, if you hear that your spouse is hanging out with someone that you're jealous of, really listen to all the information. Just because your spouse happens to be talking to that person doesn't mean it's something for you to be jealous of. She may be talking to him because she mistakenly ran into him and felt it'll be rude not to talk to him.

Love: when love does not fit the fact, the opposite action is to avoid the person that you're getting overly attached to. You want to avoid the places that you might run into the person, whether it's on social media, or at the corner market. Try to distract yourself from the thought of that person. Do something to distract yourself, whether It's gardening, cycling, or golfing; just do anything active and would distracting you away from that person that you love does not fit the fact for them. Another way to do the opposite action of love is to constantly remind yourself of the cons of loving the person. So when you find your love stirring up, think of the times that you didn't love the person or the times that the person hurt you. Do the opposite actions of the love urges. For Instance, if you find yourself buying gifts for that person, try to get rid of anything in your possession that will remind you of that person. If you are listening to a song that reminds you of the pleasant times you had with that person, instead listen to a song that reminds you of the unpleasant times you had with that person. Avoid contact with anything that makes you remember your love for that person. If you still have a strong feeling for that person, then don't say it out loud. Use unloving body facial expressions and gestures towards that person. All these may sound counter-intuitive until you start to think of the fact that love may not really be healthy. You might think that love is always great, but that may not always be the case especially if you're someone with abandonment issues, you may not be able to look at whether your relationship with someone is healthy or not because you're acting on the issues of the fear of abandonment. People with BDP normally find out that they're afraid of leaving unhealthy relationships even if it's for their own good because they're so scared of that fear of abandonment. So if you find that your relationship with that person is unhealthy for you, these tips are going to help you move out of that relationship.

Applied Behavior Analysis: Do It Yourself

Not everyone has the monetary funds or qualifies for expensive board certified, or clinically licensed professionals. I'd like to present an understandable manual for parents to deliver interventions based on ABA principals and Special Education standards that are both effective and potentially long lasting solutions to difficult behaviors surrounding any targeted behaviors. I'd like to share information in laymen's terms so everyone can both understand and implement various interventions at home, or anywhere for that matter. Living with, raising or having a relationship with someone that suffers from autism or a serious learning disability or emotional impairment can be difficult. You can help them and still make a difference in their life. Everyone deserves a chance at happiness, especially those in need and their caretakers. This manual will only contain what I perceive to be the most practical condensed version of effective at home do it yourself related interventions for parents, siblings or even spouses who wish to attempt to help others in need. This program may have been endorsed for autism, but the given strategies can be main-streamed to other disabilities or emotional impairment as well.

Antecedent Manipulation

ABA interventions use a powerful theory that ABA trained technicians call antecedent manipulation. All that this means is that you are using the environment or some other stimuli to your advantage to manipulate their behavior one way or another. When you implement this tactic try to think of yourself as a salesman. You need to use positive facial expressions, body language and verbal expression. You're providing the key ingredient to encourage or discourage a behavior. It is also similar to hanging a stick with a carrot at the end of it. Whatever you're using is positive reinforcement. You don't want to provide so much that it's not economical, nor do you want to provide so little that it's un-noticeable. If its food to be given as a positive reinforcement that's okay. Try to keep it as simple as a really healthy snack in moderation. Something that may also be economical would be a pat on the back, a smile and some verbal praise. This is also something you don't want to overdo. Facial expressions and body language is important. Those who may have a deficit in one area or more can usually read others very well, especially their caretakers, loved ones or family members. These individuals may have a deficit, but when it comes to you they are extremely intuitive. It is very important that you provide the stimuli or otherwise called manipulation as quickly as possible after the targeted behavior you want is exhibited by the individual. You want their brain to draw a relationship between how they just behaved with the idea that it was a good thing. Not only that, but you want them to enjoy the reward they received directly after that behavior. You are attempting to rewire what they define as either good or bad. There are a couple of analogies that can be used to think about one's mind. Our minds are like computers in that they are constantly processing things. With individuals that suffer from autism or potentially a severe learning disability or emotional deficit their computer processes either at a faster speed, or a lower speed than that of many others. For whatever reason their mind is going to accept, reject or not interpret information that many others adapt to. It is important to avoid assuming that the targeted individual you're working with knows what you mean. Repetition is key both relating to verbal reinforcement, explaining and also routine. You can also take something away to draw a line in someone's mind in relationship to it defining it as a good or bad thing. For example if a person attempts to avoid doing their chores through constantly telling you that their television show is on you need to then remove that stimuli. You simply need to turn the television off. If they continue to struggle it is important that you ignore their behavior and continue to be firm in your resolution surrounding them completing their chores. Negative reinforcement is taking something away in order to encourage a behavior. Some scorn upon negative reinforcement, but negative reinforcement is not always a bad thing and sometimes it is mandatory to get someone on task in relationship to what you want them to complete. When someone is training someone in relation to a skill or pretty much anything it is important to measure. Without measuring we may not have a clear cut idea in relation to whether we are being effective, or ineffective. In conventional schools there are tests and exams. Exams come at the end of the term, but tests can come up any time. They are implemented to reflect scores.

When you're measuring a behavior you need some way of keeping track of their success so you know whether they're making progress or not. One thing you need to be aware of is that individuals with severe disabilities such as autism, or other issues may not progress as fast as you are anticipating or hoping. Success takes a lot of time and you may need to step back from one targeted behavior and lower the bar. It may also behoove you to concentrate on a different targeted behavior if the one you are working on is wearing you out. Another potential idea is to sometimes just lower the bar. Expectations have to be realistic surrounding what is called the person's baseline. The baseline is basically the realm of where that person interprets, communicates, behaves and what their overall brain functioning encompasses. It is important to consider their baseline prior to any objective or goal you're attempting to work on with them. This can relate to interpersonal, hygiene, social any maladaptive behavior whatsoever.

Charting & Measuring

It is important to create a chart, diary or other instrument to document to yourself how they behaved after you attempted to implement something. You don't have to tediously write down every single behavior, but more of an overall view of what occurred. If the state was paying you for doing so you would have to do it daily and weekly. Since you are performing as your own therapist for someone else in need I suggest documenting progress at least once or twice a week. Some ideas for a charting progress can be using an excel spreadsheet, but also a simple diary. You can use resources like a computer or just a simple .99 notepad and a pencil. Measuring doesn't have to be expensive or elaborate. It can be as complex as a mathematical equation or as simple as chicken scratches or a paragraphed synopsis of what occurred during your intervention. As long as you understand it and can explain it and interpret it than you are on the right path. It also must be measurable. Using a percent such as two out of three times successful is perfect. An example of this would be, "Jim was able to answer the door two out of three times". This makes it simple, yet gives an unmistakable view as to exactly where this person is functioning. It's important to date these observations so you can look back and tell when you observed it. Then the next week if the targeted behavior decreases to one out of three times you can say it decreased and you have evidence to show that it decreased if you document it. Attempt to avoid using generic words or generalizations while you are charting the person's progress. Interpreting it later is more difficult the more general you are, but the more specific or measurable you are the more objective it will be. You don't want to write, "Jim did okay today and hopefully next week will be better". The last statement doesn't give you a clear enough of an indicator as to how Jim is doing. Generic or general terminology should be kept to a minimum. You need something mathematical somewhere in your synopsis so that later down the road when you look at it there will be no second guessing. Second guessing at least will be to a minimum with the help of measurable language such as a percent, or fraction. Fifty percent, or one out or two works is great for your record keeping.

Documentation is very important, but don't allow this chore to keep you from losing focus on the main ingredient of successful intervention, which is antecedent manipulation. You are using the art of manipulation to your advantage. In a way you are brain washing, except that your intentions are good. Brain washing is a powerful weapon that has been proven time and time again. An example of when brain washing was not implemented for the greater good was during World War II when Hitler took power. You are implementing brain washing related tools and techniques for the greater good, instead of evil. You are rewiring their mind and what they interpret or how they feel. This will in turn change how they perceive many different things. In some ways you are creating a brand new person, except they'll be stronger and more able to adapt to expectations set forth in the world. In the world of ABA therapy measuring outcomes is called discreet trials. You don't have to worry about that terminology since you are going to play the part of the therapist. You can define it as simple or as complex as you wish, but you have to do it.

Applied Behavior Analysis encourages you to somewhat break some of a targeted person's routine, while traditional Special Education related interventions encourages routine. This is where you need to decide what area you wish to target first. It is not a good idea to target everything at once because you will have chaos at a level that would be difficult to manage. Choose hygiene or a basic social skill at first. Work on 1-2 things for a while before moving onto something else. Don't expect them to get 100% on something before moving onto something else. You can abandon an objective and return to it later down the road. Once you have modified one of their behaviors you can build a new routine reinforcing that behavior. Routines help by aiding a continuation of a behavior through creating the expectation in someone's mind on what to expect relating to their environment and their actions. A simple example of a specific morning routine you may already implement could be making a hot pot of coffee every single morning. A person gets acclimated to smelling coffee and associating it with the taste of coffee and developing an idea of comfort or waking up with the ideology of making coffee in the morning. Before you know it that person is making coffee every morning. Their brain is associating coffee with waking up along with pleasurable smells.

Be Consistent & Be Patient

While you are working with the person on targeted behaviors it is both important in regards to Special Education related principals and ABA related tactics to stay consistent. If you skip a wanted or planned routine that is occurring inside the mind-set of the individual you're working with you are then decreasing the chance for them to master that behavior. For example if you want them to like coffee, but skip making coffee and give them a hot chocolate their mind may interpret coffee as no big deal. They may even interpret hot chocolate as better than coffee. I'm not encouraging you to teach anyone that coffee is good, but using coffee as an example. You are programming the person similar to reprogramming the most complex computer on the planet and consistency is important. You may at times be tired, or want to skip a planned routine surrounding what you're attempting to do. This is human, but we have to stick with the program since we are reprogramming or attempting to at least reprogram them.

Focus is important. Unless it is by design and you're measuring something you should attempt to keep the routine the same. This is interpreted as therapeutic and aids in the reduction of melt downs and potential negative behaviors in the individual you're working with. Breaking routine by design is perfectly okay. When you are doing this by design you are attempting to use it as a tool to teach the individual and potentially change a behavior. The individual you work with will be able to tell if you are focused or if you are attempting to avoid something. You need to attempt to mask your exhaustion or anything that may distract you or them from achieving whatever it is that you both are working towards. Staying positive is important and necessary for both you and them. Keeping it all together will be difficult, but possible. There isn't anything wrong with changing a routine to give yourself a break, but you don't want to completely abandon anything. You can always work on things on a lower level. Taking things down a few notches can potentially give both you and the person you're working with a much needed reduction in stress and energy.

Keeping Your Sanity

It is also important at this time to not lose focus on the big picture. If you feel exhausted and tired surrounding a particular goal you're working on with them just be patient. Some things may take longer times than others. A person's baseline doesn't change overnight and if the goal is way above their current baseline you're not going to potentially achieve 100% relating to that particular goal. From my observation surrounding obtaining success is it occurs very slowly, yet very gradually. While you are working with individuals that have autism and other issues there are fundamental principles you need to follow to keep your patience and your sanity. Give yourself time to surround yourself with positive people. Find ways to enjoy friends and family. While you are programming the targeted individual attempt to possibly program your own mind into enjoying even the most simplest of things such as sharing a cup of coffee with a friend or having a warm conversation on the phone with someone you care about. It could also be taking a hot bath with the lights off and a candle lit. It could possibly be a brief trip to the gym or making a hot fudge sundae and devour it. Whatever it is you decide to do to better your day attempt to also be consistent with that because without rewarding yourself you'll eventually lose focus and energy.

Take Baby Steps

If you are working with an individual on coping with going to the mall, instead of attempting to go to the mall take them to a secluded park. Better yet take them to a park when there isn't a lot of traffic or people around. When or if a meltdown occurs it will be less stressful and more manageable surrounding your objective. People with autism tend to get over stimulated by noise or any kind of stimulation. When an individual becomes over stimulated they are going to be at a higher stress level, which will cause a potential break down. You will experience break downs in individuals while you are attempting to increase targeted behaviors however if you choose the right settings the stress from these break downs is at least minimized. If you are toilet training a puppy you start with puppy pads. A baby must learn to crawl before it can walk. All things will evolve in a chronological and systematic way if you allow them to do so. Processing information takes a longer amount of time if that information is brand new to one's brain. While one is learning new information synapses are formed inside the brain. When or if a stressful event takes place it impedes the formation of those synapses. When events occur attempt to down scale them to an extent. Handle the misbehavior, but don't create a new stressor in the mind of the one you're attempting to help.

Concentrate On the Positives

While focusing your energy and direction towards an individual obtaining or learning new skills you'll find that they're going to grasp certain things much easier than others. It is important to focus on the positives they exhibit. They will fall down much more than they will be able to get up, especially while processing new information. Technically a therapist will move on once a certain skill is mastered, however if you don't have the energy to move on yourself it is okay to attempt to work on some of the same skill sets. If it replicates the routine or the therapy you're providing and you aren't ready yourself to move on at least you are continuing the process. You may need to also focus on the positives surrounding your own strengths and weaknesses. If you know you're particularly fond of a particular activity with that individual than implement that even when you are done teaching that skill. If it helps you when you're exhausted it is absolutely alright to do so. If it were a paid therapist they may do the same and document and identify it as the maintenance of a skill. You can simply justify it under that umbrella.

Finding Supports

There are support groups for those who aid in the care of others with autism, or severe emotional impairment as well. You can obtain information in regards to those support groups from such resources as your local Health Department, or your local library. Either may have information that you need. Many times there is a section in the newspaper that specifically lists support groups. It is okay to utilize your faith and your potential clergy in regards to obtaining the support you need. We are all human and needing some support every now and then is perfectly normal.

Closure

The strategies that have been shared related to antecedent manipulation, positive/negative reinforcement, consistency, measuring with objectivity and strategies to attempt to chart that progress. This brief manual is a tool for those that wish to attempt to aid in the skill progression of others in need. Those are some of the strategies have been identified for autism for which they fall under the umbrella of Special Needs. They can be used however you deem necessary.

Conclusion

Dialectical Behavior Therapy provides an ample of benefits. The four modules (mindfulness, interpersonal effectiveness, emotional regulation, and distress tolerance) are designed to assist people with depression, anxiety, borderline personality disorder, eating disorder and much more.

Each module is interrelated to one another, and learning each component is essential to reaping the benefits of DBT.

It is the innate desire of this eBook that you acquire a greater understanding of DBT and its functions.

Remember: Life is not handed to us on a silver platter; everyone in this world is struggling with a specific issue—you're not alone.